Looking
for Spinoza

ALSO BY ANTONIO DAMASIO

The Feeling of What Happens
Descartes' Error

ANTONIO DAMASIO

Looking
for Spinoza

*Joy, Sorrow,
and the
Feeling Brain*

HARCOURT, INC.

Orlando   Austin   New York   San Diego   Toronto   London

www.HarcourtBooks.com

Library of Congress Cataloging-in-Publication Data
Damasio, Antonio R.
Looking for Spinoza: joy, sorrow, and the feeling brain/
Antonio Damasio.—1st ed.
p.   cm.
Includes bibliographical references and index.
ISBN 0-15-100557-5
1. Emotions. 2. Neuropsychology. 3. Spinoza, Benedictus de, 1632–1677.
I. Title.
QP401 .D203 2003
152.4—d21   2002011347

All figures, diagrams, and drawings are by Hanna Damasio except for the
portrait on page 263. Her drawings in Chapters 1, 5, and 6 depict Spinoza's
house on 72-74 Paviljoensgracht (page 9), a statue of Spinoza (page 16), the
back of the New Church and Spinoza's tomb in The Hague (page 19), the
Portuguese Synagogue in Amsterdam (page 185), the house where Spinoza
lived in Rijnsburg (page 223), a bust of Spinoza (page 225), and the old
synagogue in Amsterdam (inspired by a 17th century engraving by Jan
Veenhuysen). The portrait on page 263 is by Jean Charles François and was
published by A. Savérien, *Histoire des Philosophes Modernes*, Paris, 1761.

Text set in Scala
Designed by Linda Lockowitz

Printed in the United States of America

First edition
K J I H G F E D C B A

To Hanna

Contents

CHAPTER 5   Body, Brain, and Mind

CHAPTER 6   A Visit to Spinoza

Looking
for Spinoza

CHAPTER I

Enter Feelings

Enter Feelings

Feelings of pain or pleasure or some quality in between are the bedrock of our minds. We often fail to notice this simple reality because the mental images of the objects and events that surround us, along with the images of the words and sentences that describe them, use up so much of our overburdened attention. But there they are, feelings of myriad emotions and related states, the continuous musical line of our minds, the unstoppable humming of the most universal of melodies that only dies down when we go to sleep, a humming that turns into all-out singing when we are occupied by joy, or a mournful requiem when sorrow takes over.*

Given the ubiquity of feelings, one would have thought that their science would have been elucidated long ago—what feelings are, how they work, what they mean—but that is hardly the case. Of all the mental phenomena we can describe, feelings and their essential ingredients—pain and pleasure—are the least understood in biological and specifically neurobiological terms. This is all the more puzzling considering that advanced societies cultivate

---

*The principal meaning of the word feeling refers to some variant of the experience of pain or pleasure as it occurs in emotions and related phenomena; another frequent meaning refers to experiences such as touch as when we appreciate the shape or texture of an object. Throughout this book, unless otherwise specified, the term feeling is always used in its principal meaning.

feelings shamelessly and dedicate so many resources and efforts to manipulating those feelings with alcohol, drugs of abuse, medical drugs, food, real sex, virtual sex, all manner of feel-good consumption, and all manner of feel-good social and religious practices. We doctor our feelings with pills, drinks, health spas, workouts, and spiritual exercises, but neither the public nor science have yet come to grips with what feelings are, biologically speaking.

I am not really surprised at this state of affairs, considering what I grew up believing about feelings. Most of it simply was not true. For example, I thought that feelings were impossible to define with specificity, unlike objects you could see, hear, or touch. Unlike those concrete entities, feelings were intangible. When I started musing about how the brain managed to create the mind, I accepted the established advice that feelings were out of the scientific picture. One could study how the brain makes us move. One could study sensory processes, visual and otherwise, and understand how thoughts are put together. One could study how the brain learns and memorizes thoughts. One could even study the emotional reactions with which we respond to varied objects and events. But feelings—which can be distinguished from emotions, as we shall see in the next chapter—remained elusive. Feelings were to stay forever mysterious. They were private and inaccessible. It was not possible to explain how feelings happened or where they happened. One simply could not get "behind" feelings.

As was the case with consciousness, feelings were beyond the bounds of science, thrown outside the door not just by the naysayers who worry that anything mental might actually be explained by neuroscience, but by card-carrying neuroscientists themselves, proclaiming allegedly insurmountable limitations. My own willingness to accept this belief as fact is evidenced by the many years I spent studying anything but feelings. It took me awhile to see the degree to which the injunction was unjustified and to realize that the neurobiology of feelings was no less viable than the neu-robiology of vision or memory. But eventually I did, mostly, as it turns out, because I was confronted by the reality of neurological patients whose symptoms literally forced me to investigate their conditions.

Imagine, for example, meeting someone who, as a result of damage to a certain location of his brain, became unable to feel compassion or embarrassment—when compassion or embar-rassment were due—yet could feel happy, or sad, or fearful just as normally as before brain disease had set in. Would that not give you pause? Or picture a person who, as a result of damage located elsewhere in the brain, became unable to experience fear when fear was the appropriate reaction to the situation and yet still could feel compassion. The cruelty of neurological disease may be a bottomless pit for its victims—the patients and those of us who are called to watch. But the scalpel of disease also is responsible for its single redeeming feature: By teasing apart the normal op-erations of the human brain, often with uncanny precision, neu-rological disease provides a unique entry into the fortified citadel of the human brain and mind.

Reflection on the situation of these patients and of others with comparable conditions raised intriguing hypotheses. First, individual feelings could be prevented through damage to a dis-crete part of the brain; the loss of a specific sector of brain cir-cuitry brought with it the loss of a specific kind of mental event. Second, it seemed clear that different brain systems controlled different feelings; damage to one area of the brain anatomy did not cause all types of feelings to disappear at once. Third, and most surprising, when patients lost the ability to express a cer-tain emotion, they also lost the ability to experience the corre-sponding feeling. But the opposite was not true: Some patients who lost their ability to experience certain feelings still could express the corresponding emotions. Could it be that while emotion and feeling were twins, emotion was born first and feeling second, with feeling forever following emotion like a shadow? In spite of their close kinship and seeming simultaneity, it seemed that emotion preceded feeling. Knowledge of this specific relationship, as we shall see, provided a window into the investigation of feelings.

Such hypotheses could be tested with the help of scanning techniques that allow us to create images of the anatomy and activity of the human brain. Step by step, initially in patients and then in both patients and people without neurological disease, my colleagues and I began to map the geography of the feeling brain. We aimed at elucidating the web of mechanisms that allow our thoughts to trigger emotional states and engender feelings.[1]

Emotion and feeling played an important but very different part in two of my previous books. *Descartes' Error* addressed the role of emotion and feeling in decision-making. *The Feeling of What Happens* outlined the role of emotion and feeling in the construction of the self. In the present book, however, the focus is on feelings themselves, what they are and what they provide. Most of the evidence I discuss was not available when I wrote the previous books, and a more solid platform for the understanding of feelings has now emerged. The main purpose of this book, then, is to present a progress report on the nature and human significance of feelings and related phenomena, as I see them now, as neurologist, neuroscientist, and regular user.

The gist of my current view is that feelings are the expression of human flourishing or human distress, as they occur in mind and body. Feelings are not a mere decoration added on to the emotions, something one might keep or discard. Feelings can be and often are *revelations* of the state of life within the entire organism—a lifting of the veil in the literal sense of the term. Life being a high-wire act, most feelings are expressions of the struggle for balance, ideas of the exquisite adjustments and corrections without which, one mistake too many, the whole act collapses. If anything in our existence can be revelatory of our simultaneous smallness and greatness, feelings are.

How that revelation comes to mind is itself beginning to be revealed. The brain uses a number of dedicated regions working in concert to portray myriad aspects of the body's activities in the form of neural maps. This portrait is a composite, an ever-changing picture of life on the fly. The chemical and neural channels that bring into the brain the signals with which this life portrait can be painted are just as dedicated as the canvas that receives them. The mystery of how we feel is a little less mysterious now.

It is reasonable to wonder if the attempt to understand feelings is of any value beyond the satisfaction of one's curiosity. For a number of reasons, I believe it is. Elucidating the neurobiology of feelings and their antecedent emotions contributes to our views on the mind-body problem, a problem central to the understanding of who we are. Emotion and related reactions are aligned with the body, feelings with the mind. The investigation of how thoughts trigger emotions and of how bodily emotions become the kind of thoughts we call feelings provides a privileged view into mind and body, the overtly disparate manifestations of a single and seamlessly interwoven human organism.

The effort has more practical payoffs, however. Explaining the biology of feelings and their closely related emotions is likely to contribute to the effective treatment of some major causes of human suffering, among them depression, pain, and drug addiction. Moreover, understanding what feelings are, how they work, and what they mean is indispensable to the future construction of a view of human beings more accurate than the one currently available, a view that would take into account advances in the social sciences, cognitive science, and biology. Why is such a construction of any practical use? Because the success or failure of humanity depends in large measure on how the public and the institutions charged with the governance of public life incorporate that revised view of human beings in principles and policies. An understanding of the neurobiology of emotion and feelings is a key to the formulation of principles and policies capable of reducing human distress and enhancing human flourishing. In effect, the new knowledge even speaks to the manner in which humans deal with unresolved tensions between sacred and secular interpretations of their own existence.

Now that I have sketched my main purpose, it is time to explain why a book dedicated to new ideas on the nature and significance of human feeling should invoke Spinoza in the title. Since I am not a philosopher and this book is not about Spinoza's philosophy, it is sensible to ask: why Spinoza? The short explanation is that Spinoza is thoroughly relevant to any discussion of human emotion and feeling. Spinoza saw drives, motivations, emotions, and feelings—an ensemble Spinoza called *affects*—as a central aspect of humanity. Joy and sorrow were two prominent concepts in his attempt to comprehend human beings and suggest ways in which their lives could be lived better.

The long explanation is more personal.

The Hague

December 1, 1999. The friendly doorman of the Hotel des Indes insists: "You should not walk in this weather, sir, let me get a car for you. The wind is bad. It is almost a hurricane, sir. Look at the flags." True, the flags have taken wing, and the fast-moving clouds are racing toward the east. Although The Hague's Embassy Row seems about to lift off, I decline the offer. I prefer to walk, I say. I will be all right. Besides, see how beautiful the sky looks in between the clouds? My doorman has no idea where I am going, and I am not going to tell him. What would he have thought?

The rain has almost stopped and with some determination it is easy to overcome the wind. I actually can walk fast and follow my mental map of the place. At the end of the promenade in front of the Hotel des Indes, to my right, I can see the old palace and the Mauritshuis, festooned with Rembrandt's face—they are showing a retrospective of his self-portraits. Past the museum square the streets are almost deserted, although this is the center of town and it is a regular working day. There must be warnings telling people to stay indoors. So much the better. I arrive at the Spui without having to brave a crowd. After I get to the New Church, the route is entirely unfamiliar and I hesitate for a second, but the choice becomes clear: I turn right on Jacobstraat, then left on Wagenstraat, then right again on Stilleverkade. Five minutes later I am on the Paviljoensgracht. I stop in front of number 72–74.

The front of the house is much as I imagined it, a small build-ing with three floors, three windows wide, a version of the average canal townhouse, more modest than  rich. It is well kept and not very differ-ent from what it must have looked like in the seventeenth century. All the win-dows are closed, and there is no sign of activity. The door is well kept and well painted, and next to it there is a shiny brass bell, set in the frame. The word SPINOZAHUIS is etched in the rim. I press the button resolutely but without much hope. There is no sound from inside and no movement in any curtain. No one had answered the phone when I tried to call earlier. Spinoza is closed for business.

This is where Spinoza lived the last seven years of his brief life and where he died in 1677. The *Theologico-Political Treatise,* which he carried when he arrived, was published from here, anonymously. The *Ethics* was completed here and published after his death, almost as anonymously.

I have no hope of seeing the house today but all is not lost. In the landscaped middle section that separates the two lanes of the street, an unexpected urban garden, I discover Spinoza himself, semiobscured by the windswept foliage, sitting quietly and pensively, in sturdy bronze perpetuity. He looks pleased and entirely undisturbed by the meteorological commotion, as well he should, having survived stronger forces in his day.

For the past few years I have been looking for Spinoza, sometimes in books, sometimes in places, and that is why I am here today. A curious pastime, as you can see, and one that I had never planned to adopt. The reason why I did has a lot to do with coincidence. I first read Spinoza as an adolescent—there is no better age to read Spinoza on religion and politics—but it is fair to say that while some ideas made lasting impressions, the reverence I developed for Spinoza was rather abstract. He was both fascinating and forbidding. Later, I never thought of Spinoza as especially relevant to my work, and my acquaintance with his ideas was sparse. And yet there was a quote of his that I had long treasured—it came from the *Ethics* and pertained to the notion of self—and it was when I thought of citing it and needed to check its accuracy and context that Spinoza returned to my life. I found the quote, all right, and it did match the contents of the yellowed paper I had once pinned to a wall. But then I started reading backward and forward from the particular passage where I had landed, and I simply could not stop. Spinoza was still the same, but I was not. Much of what once seemed impenetrable now seemed familiar, strangely familiar, in fact, and quite relevant to several aspects of my recent work. I was not about to endorse all of Spinoza. For one thing, some passages were still opaque, and there were conflicts and inconsistencies of ideas unresolved after multiple readings. I still was puzzled and even exasperated. Mostly, however, for better or worse, I found myself in a pleasant resonance with the ideas, a bit like the character in Bernard Malamud's *The Fixer*, who read a few pages of Spinoza and who kept on going as though there were a whirlwind on his back: "... I didn't understand every word but when you're dealing with such ideas you feel as though you were taking a witch's ride."[2] Spinoza dealt with the subjects that preoccupy me most as a scientist—the nature of emotions and feelings and the relation of mind to body—and those same subjects have preoccupied many other thinkers of the past. To my eyes, however, he seemed to have prefigured solutions that researchers are now offering on a number of these issues. That was surprising.

For example, when Spinoza said that *love is nothing but a pleasurable state, joy, accompanied by the idea of an external cause,* he was separating with great clarity the process of feeling from the process of having an idea about an object that can cause an emotion.[3] Joy was one thing; the object that caused joy was another. Joy or sorrow, along with the idea of the objects that caused either, eventually came together in the mind, of course, but they were distinct processes to begin with, within our organisms. Spinoza had described a functional arrangement that modern science is revealing as fact: Living organisms are designed with an ability to react emotionally to different objects and events. The reaction is followed by some pattern of feeling and a variation of pleasure or pain is a necessary component of feeling.

Spinoza also proposed that the power of affects is such that the only hope of overcoming a detrimental affect—an irrational passion—is by overpowering it with a stronger positive affect, one triggered by reason. *An affect cannot be restrained or neutralized except by a contrary affect that is stronger than the affect to be restrained.*[4] In other words, Spinoza recommended that we fight a negative emotion with an even stronger but positive emotion brought about by reasoning and intellectual effort. Central to his thinking was the notion that the subduing of the passions should be accomplished by reason-induced emotion and not by pure reason alone. This is by no means easy to achieve, but Spinoza saw little merit in anything easy.

Of great importance for what I shall be discussing was his notion that both the mind and the body were parallel attributes (call them manifestations) of the very same substance.[5] At the very least, by refusing to ground mind and body on different substances, Spinoza was serving notice of his opposition to the view of the mind-body problem that prevailed in his time. His dissent stood out in a sea of conformity. More intriguing, however, was his notion that *the human mind is the idea of the human body.*[6] This raised an arresting possibility. Spinoza might have intuited the principles behind the natural mechanisms responsible for the parallel manifestations of mind and body. As I shall discuss later, I am convinced that mental processes are grounded in the brain's mappings of the body, collections of neural patterns that portray responses to events that cause emotions and feelings. Nothing could have been more comforting than coming across this statement of Spinoza's and wondering about its possible meaning.

This would have been more than enough to fuel my curiosity about Spinoza, but there was more to sustain my interest. For Spinoza, organisms naturally endeavor, of necessity, to persevere in their own being; that necessary endeavor constitutes their actual essence. Organisms come to being with the capacity to regulate life and thereby permit survival. Just as naturally, organisms strive to achieve a "greater perfection" of function, which Spinoza equates with joy. All of these endeavors and tendencies are engaged unconsciously.

Darkly, through the glass of his unsentimental and unvarnished sentences, Spinoza apparently had gleaned an architecture of life regulation along the lines that William James, Claude Bernard, and Sigmund Freud would pursue two centuries later. Moreover, by refusing to recognize a purposeful design in nature, and by conceiving of bodies and minds as made up of components that could be combined in varied patterns across different species, Spinoza was compatible with Charles Darwin's evolutionary thinking.

Armed with this revised conception of human nature, Spinoza proceeded to connect the notions of good and evil, of freedom and salvation, to the affects and to the regulation of life. Spinoza suggested that the norms that govern our social and personal conduct should be shaped by a deeper knowledge of humanity, one that made contact with the God or Nature *within* ourselves.

Some of Spinoza's ideas are part and parcel of our culture, but to the best of my knowledge Spinoza is absent as a reference from the modern efforts to understand the biology of the mind.[7] This absence is interesting in itself. Spinoza is a thinker far more famous than known. Sometimes Spinoza appears to rise out of nothing, in solitary and unexplained splendor, although the impression is false—in spite of his originality he is very much a part of his intellectual times. And he appears to dissolve as abruptly, without succession—another false impression given that the essence of some of his forbidden proposals can be found behind the Enlightenment and well beyond in the century that followed his death.[8]

One explanation for Spinoza's status as unknown celebrity is the scandal he caused in his own time. As we shall see (in Chapter Six), his words were deemed heretical and banned for decades and with rare exceptions were quoted only as part of the assault on his work. The attacks paralyzed most attempts by Spinoza admirers to discuss his ideas publicly. The natural continuity of intellectual acknowledgment that follows a thinker's work was thus interrupted, even as some of his ideas were used uncredited. This state of affairs, however, hardly explains why Spinoza continued to gain fame but remained unknown once the likes of Goethe and Wordsworth began to champion him. Perhaps a better explanation is that Spinoza is not easy to know.

The difficulty begins with the problem that there are several Spinozas with which to reckon, at least four by my count. The first is the accessible Spinoza, the radical religious scholar who disagrees with the churches of his time, presents a new conception of God, and proposes a new road to human salvation. Next comes Spinoza as political architect, the thinker who describes the traits of an ideal democratic state populated by responsible, happy citizens. The third Spinoza is the least accessible of the set: the philosopher who uses scientific facts, a method of geometric demonstration and intuition to formulate a conception of the universe and the human beings in it.

Recognizing these three Spinozas and their web of dependencies is enough to suggest how convoluted Spinoza can be. But there is a fourth Spinoza: the protobiologist. This is the biological thinker concealed behind countless propositions, axioms, proofs, lemmas, and scholia. Given that many of the advances on the science of emotions and feeling are consonant with proposals that Spinoza began to articulate, my second purpose in this book is to connect this least-known Spinoza to some of the corresponding neurobiology of today. But I note, again, that this book is not about Spinoza's philosophy. I do not address Spinoza's thinking outside of the aspects I regard as pertinent to biology. The goal is more modest. One of the values of philosophy is that throughout its history it has prefigured science. In turn, I believe, science is well served by recognizing that historical effort.

Looking for Spinoza

Spinoza is relevant to neurobiology in spite of the fact that his reflections on the human mind came out of a larger concern for the condition of human beings. Spinoza's ultimate preoccupation was the relation of human beings to nature. He attempted to clarify that relationship so he could propose realistic means for human salvation. Some of those means were personal, under the sole control of the individual, and some relied on the help that certain forms of social and political organization provided the individual. His thinking descends from Aristotle's, but the biological grounding, not surprisingly, is firmer. Spinoza seems to have gleaned a relation between personal and collective happiness, on the one hand, and human salvation and the structure of the state, on the other, long before John Stuart Mill. At least regarding the social consequences of his thinking there seems to be considerable recognition.[9]

Spinoza prescribed an ideal democratic state, where the hallmarks were freedom of speech—*let every man think what he wants and say what he thinks,* he wrote[10]—separation of church and state, and a generous social contract that promoted the well-being of citizens and the harmony of government. Spinoza offered this prescription more than a century ahead of the Declaration of Independence and First Amendment. That Spinoza, as a part of his revolutionary endeavors, also anticipated some aspects of modern biology is all the more intriguing.

Who was this man, then, who could think about mind and body in ways that were not only profoundly opposed to the thinking of most of his contemporaries, but remarkably current three hundred and some years later? What circumstances produced such a contrary spirit? To attempt an answer to these questions, we must consider yet another Spinoza, the man behind three distinct first names—Bento, Baruch, Benedictus—a person at once courageous and cautious, uncompromising and accommodating, arrogant and modest, detached and gentle, admirable and irritating, close to the observable and the concrete and yet unabashedly spiritual. His personal feelings are never revealed directly in his writings, not even in his style, and he must be pieced together from a thousand indirections.

Almost without noticing, I began looking for the person behind the strangeness of the work. I simply wanted to meet the man in my imagination and chat a little, have him sign *The Ethics* for me. Reporting on my search for Spinoza and the story of his life became the third purpose of this book.

Spinoza was born in the prosperous city of Amsterdam in 1632, literally in the middle of Holland's Golden Age. That same year, a brief walk from the Spinoza household, a twenty-three-year-

old Rembrandt van Rijn was painting *The Anatomy Lesson of Dr. Tulp,* the picture that began his fame. Rembrandt's patron, Constantijn Huygens, statesman and poet, secretary to the Prince of Orange, and friend of John Donne, had recently become the father of Christiaan Huygens, who was to be one of the most celebrated astronomers and physicists of all time. Descartes, the leading philosopher of the day, then thirty-two, also was living in Amsterdam, on the Prinsengraacht, and worrying about how his new ideas on human nature would be received in Holland and abroad. Soon he would come to teach algebra to young Christiaan Huygens. Spinoza came into the world amid embarrassing riches, intellectual and financial, to draw on Simon Schama's apt descriptor of the place in this age.[11]

Bento was the name Spinoza received at his birth from his parents, Miguel and Hana Debora, Portuguese Sephardic Jews who had resettled in Amsterdam. He was known as Baruch in the synagogue and among friends while he was growing up in Amsterdam's affluent community of Jewish merchants and scholars. He adopted the name Benedictus at age twenty-four after he was banished by the synagogue. Spinoza abandoned the comfort of his Amsterdam family home and began the calm and deliberate errancy whose last stop was here in the Paviljoensgracht. The Portuguese name Bento, the Hebrew name Baruch, and the Latin name Benedictus, all mean the same: blessed. So, what's in a name? Quite a lot, I would say. The words may be superficially equivalent, but the concept behind each of them was dramatically different.

Beware

I need to get inside the house, I think, but for now the door is closed. All I can do is imagine someone emerging from a barge moored close to it, walking into the house, and inquiring after Spinoza (the Paviljoensgracht was a wide canal, in those days; later it was filled in and turned into a street, as were so many canals in Amsterdam and Venice). The wonderful Van der Spijk, the owner and a painter, would open the door. He would amiably usher the visitor into his studio, behind the two windows next to the main door, invite him to wait, and go tell Spinoza, his lodger, that a caller had arrived.

Spinoza's rooms were on the third floor, and he would come down the spiral staircase, one of those tightly curled, horrifying

stairs for which Dutch architecture is infamous. Spinoza would
be elegantly dressed in his *fidalgo* garb—nothing new, nothing
very worn, all well kept, a white starched collar, black breeches, a
black leather vest, a black camel-hair jacket nicely balanced on his
shoulders, shiny black leather shoes with large silver buckles, and
a wood cane, perhaps, to help negotiate the stairs. Spinoza had a
thing for black leather shoes. Spinoza's harmonious and clean-
shaven face, his large black eyes shining brilliantly, would dominate his appearance. His hair was black too, as were the long
eyebrows; the skin was olive; the stature medium; the frame light.

With politeness, even affability, but with economic direct-
ness, the visitor would be prompted to come to the matter at
hand. This generous teacher could entertain discussions of optics, politics, and religious faith during his office hours. Tea would
be served. Van der Spijk would continue painting, mostly silently,
but with salubrious democratic dignity. His seven ebullient children would stay out of the way in the back of the house. Mrs. Van
der Spijk sewed. The help toiled away in the kitchen. You see the
picture.

Spinoza would be smoking his pipe. The aroma would do
battle with the fragrance of turpentine as questions were pon-
dered, answers given, and daylight waned. Spinoza received countless visitors, from neighbors and relatives of the Van der Spijks to
eager young male students and impressionable young women,
from Gottfried Leibniz and Christiaan Huygens to Henry Olden-
burg, president of the newly created Royal Society of Britain.
Judging from the tone of his correspondence he was most charitable with the simple folk and least patient with his peers. Apparently he could suffer modest fools easily but not the other kind.

I also can imagine a funeral cortege, on another gray day, February 25, 1677, Spinoza's simple coffin, followed by the Van der
Spijk family, and "many illustrious men, six carriages in all,"

marching slowly to the New Church, just minutes away. I walk back to the New Church retracing their likely route. I know Spinoza's grave is in the churchyard, and from the house of the living I may as well go to the house of the dead.

Gates surround the churchyard but they are wide open. There is no cemetery to speak of, only shrubs and grass and moss and muddy lanes amid the tall trees. I find the grave much where I thought it would be, in the back part of the yard, behind the church, to the south and east, a flat stone at ground level and a vertical tombstone, weathered and unadorned. Besides announcing whose grave it is, the inscription reads CAUTE! which is Latin for "Be careful!" This is a chilling bit of advice considering Spinoza's remains are not really inside the tomb, and that his body was stolen, no one knows by whom, sometime after the burial when the corpse lay  inside the church. Spinoza had told us that every man should think what he wants and say what he thinks, but not so fast, not quite yet. Be careful. Watch out for what you say (and write) or not even your bones will escape.

Spinoza used *caute* in his correspondence, printed just beneath the drawing of a rose. For the last decade of his life his written words were indeed *sub-rosa*. He listed a fictitious printer for the *Tractatus*, along with an incorrect city of publication (Hamburg). The author's page was blank. Even so, and even though the book was written in Latin rather than Dutch, authorities in Holland prohibited it in 1674. Predictably, it also was placed in the

Vatican's Index of dangerous books. The church considered the book an all-out assault on organized religion and the political power structure. After that Spinoza refrained from publishing altogether. No surprise. His last writings still were in the drawer of his desk on the day of his death, but Van der Spijk knew what to do: He shipped the entire desk aboard a barge to Amsterdam where it was delivered to Spinoza's real publisher, John Rieuwertz. The collection of posthumous manuscripts—the much-revised *Ethics,* a *Hebrew Grammar,* the second and unfinished *Political Treatise,* and the *Essay on the Improvement of the Understanding*—was published later that same year, anonymously. We should keep this situation in mind when we describe the Dutch provinces as the haven of intellectual tolerance. Without a doubt they were, but the tolerance had its limits.

For most of Spinoza's life Holland was a republic, and during Spinoza's mature years the Grand Pensionary Jan De Witt dominated political life. De Witt was ambitious and autocratic but also was enlightened. It is not clear how well he knew Spinoza, but he certainly knew of Spinoza and probably helped contain the ire of the more conservative Calvinist politicians when the *Tractatus* began to cause scandal. De Witt owned a copy of the book since 1670. He is rumored to have sought the philosopher's opinion on political and religious matters, and Spinoza is rumored to have been pleased by the esteem De Witt showed him. Even if the rumors are untrue, there is little question De Witt was interested in Spinoza's political thinking and at least sympathetic to his religious views. Spinoza felt justifiably protected by De Witt's presence.

Spinoza's sense of relative safety came to an abrupt close in 1672 during one of the darkest hours of Holland's golden age. In a sudden turn of events, of the sort that define this politically volatile era, De Witt and his brother were assassinated by a mob, on the false suspicion that they were traitors to the Dutch cause in the ongoing war with France. Assailants clubbed and knifed both

De Witts as they dragged them on the way to the gallows, and by the time they arrived there was no need to hang them anymore. They proceeded to undress the corpses, suspend them upside down, butcher-shop style, and quarter them. The fragments were sold as souvenirs, eaten raw, or eaten cooked, amid the most sickening merriment. All this took place not far from where I am standing now, literally around the corner from Spinoza's home, and it was probably Spinoza's darkest hour as well. The attacks shocked many thinkers and politicians of the time. Leibniz was horrified and so was the unflappable Huygens, in the safety of Paris. But Spinoza was undone. The savagery revealed human nature at its shameful worst and jolted him out of the equanimity he had worked so hard to maintain. He prepared a placard that read *ULTIMI BARBORORUM* (Ultimate barbarians) and wanted to post it near the remains. Fortunately Van der Spijk's dependable wisdom prevailed. He simply locked the door and kept the key, and Spinoza was thus prevented from leaving the house and facing a certain death. Spinoza cried publicly—the only time, it is said, that others saw him in the throes of uncontrolled emotion. The intellectual safe harbor, such as it was, had come to an end.

I look at Spinoza's grave one more time and am reminded of the inscription Descartes prepared for his own tombstone: "He who hid well, lived well."[12] Only twenty-seven years separate the death of these two part-time contemporaries (Descartes died in 1650). Both spent most of their lives in the Dutch paradise, Spinoza by birthright, the other by choice—Descartes had decided early in his career that his ideas were likely to clash with the Catholic Church and monarchy in his native France and left quietly for Holland. Yet both had to hide and pretend, and in the case of Descartes, perhaps distort his own thinking. The reason should be clear. In 1633, one year after Spinoza's birth, Galileo was questioned by the Roman Inquisition and placed under house arrest. That same year Descartes withheld publication of his *Treatise of*

Man and, even so, had to respond to vehement attacks on his views of human nature. By 1642, in contradiction with his earlier thinking, Descartes was postulating an immortal soul separate from the perishable body, perhaps as a preemptive measure to forestall further attacks. If that was his intent, the strategy eventually worked, but not quite in his lifetime. Later he made his way to Sweden to mentor the spectacularly irreverent Queen Christina. He died midway through his first winter in Stockholm, at age fifty-four. Amid the thanks we must give for living in different times, even today one shudders to think of the threats against such hard-won freedoms. Perhaps *caute* still is in order.

As I leave the churchyard, my thoughts turn to the bizarre significance of this burial site. Why is Spinoza, who was born a Jew, buried next to this powerful Protestant church? The answer is as complicated as anything else having to do with Spinoza. He is buried here, perhaps, because having been expelled by his fellow Jews he could be seen as Christian by default; he certainly could not have been buried in the Jewish cemetery at Ouderkerk. But he is *not* really here, perhaps, because he never became a proper Christian, Protestant or Catholic, and in the eyes of many he was an atheist. And how fitting it all is. Spinoza's God was neither Jewish nor Christian. Spinoza's God was everywhere, could not be spoken to, did not respond if prayed to, was very much in every particle of the universe, without beginning and without end. Buried and unburied, Jewish and not, Portuguese but not really, Dutch but not quite, Spinoza belonged nowhere and everywhere.

Back at the Hotel des Indes the doorman is glad to see me in one piece. I can't resist. I do tell him that I am looking for Spinoza, that I have been to his house. The solid Dutchman is taken aback. He stops in bewilderment and utters, after a pause, "You mean . . .

the philosopher?" Well, he does know who Spinoza was, after all, Holland being one of the best educated places on earth. But he has no idea that Spinoza lived the last part of his life in The Hague, finished his most important work here, died here, is buried here—well, sort of—and has a house and a statue and a tomb to his credit here, a mere twelve blocks away. To be fair, few people have any idea of this either. "They don't speak much of him, these days," says my friendly doorman.

In the Paviljoensgracht

Two days later I return to 72 Paviljoensgracht, and this time my gracious hosts have arranged for me to visit the house. The weather is even worse today and something like a hurricane has been blowing in from the North Sea.

Van der Spijk's studio is only marginally warmer and certainly darker than outside. A mush of gray and green remains in my mind. It is a small space, easy to commit to memory, and easy to play with in one's imagination. Mentally, I rearrange the furniture, relight the room, and warm it up. I sit long enough to imagine the movements of Spinoza and Van der Spijk on this confined stage, and conclude that no amount of redecoration will turn the room into the comfortable salon that Spinoza deserved. It is a lesson in modesty. In this small space Spinoza received his countless visitors, Leibniz and Huygens included. In this small space Spinoza took his meals—when he was not too distracted with his work and forgot all about eating—and talked to Van der Spijk's wife and to their noisy children. In this small space he sat crushed by the news of the De Witts' assassination.

How could Spinoza have survived this confinement? No doubt by freeing himself in the infinite expanse of his mind, a place larger and no less refined than Versailles and its gardens, where, on those very same days, Louis XIV, barely six years younger than Spinoza and destined to survive him by another thirty, would be strolling with his large retinue in tow.

It must be that Emily Dickinson was right, that one single brain, being wider than the sky, can comfortably accommodate a good man's intellect and the whole world besides.

CHAPTER 2

Of Appetites
and Emotions

Trust Shakespeare

Trust Shakespeare to have been there before. Toward the end of *Richard II,* the crown now lost and the prospect of jail looming close, Richard unwittingly tells Bolingbroke about a possible distinction between the notion of emotion and that of feeling.[1] He asks for a looking glass, confronts his face, and studies the spectacle of ravage. Then he notes that the "external manner of laments" expressed in his face is merely "shadows of the unseen grief," a grief that "swells with silence in the tortured soul." His grief, as he says, "lies all within." In just four lines of verse, Shakespeare announces that the unified and apparently singular process of affect, which we often designate casually and indifferently as emotion *or* feeling, can be analyzed in parts.

My strategy for elucidating feelings capitalizes on this distinction. It is true that the common usage of the word emotion tends to encompass the notion of feeling. But in our attempt to understand the complex chain of events that begins with emotion and ends up in feeling, we can be helped by a principled separation between the part of the process that is made public and the part that remains private. For the purposes of my work I call the former part *emotion* and the latter part *feeling* in keeping with the meaning of the term feeling I outlined earlier. I ask the reader to accompany me in this choice of words and concepts for the good reason that it may permit us to uncover something about the biology that lies beneath. By the end of chapter 3, I promise to put emotion and feeling back together again.[2]

In the context of this book then, emotions are actions or movements, many of them public, visible to others as they occur in the face, in the voice, in specific behaviors. To be sure, some components of the emotion process are not visible to the naked eye but can be made "visible" with current scientific probes such as hormonal assays and electrophysiological wave patterns. Feelings, on the other hand, are always hidden, like all mental images necessarily are, unseen to anyone other than their rightful owner, the most private property of the organism in whose brain they occur.

Emotions play out in the theater of the body. Feelings play out in the theater of the mind.[3] As we shall see, emotions and the host of related reactions that underlie them are part of the basic mechanisms of life regulation; feelings also contribute to life regulation, but at a higher level. Emotions and related reactions seem to precede feelings in the history of life. Emotions and related phenomena are the foundation for feelings, the mental events that form the bedrock of our minds and whose nature we wish to elucidate.

Emotions and feelings are so intimately related along a continuous process that we tend to think of them, understandably, as one single thing. In the normal situation, however, we can glean different segments along the continuous process and, under the microscope of cognitive neuroscience, it is legitimate to dissociate one segment from the other. With naked eyes and a slew of scientific probes, an observer may objectively examine the behaviors that make up an emotion. In effect, the prelude to the process of feeling can be studied. Turning emotion and feeling into separate research objects helps us discover how it is that we feel.

The goal of this chapter is to explain the brain and body mechanisms responsible for triggering and executing an emotion. The focus here is on the intrinsic "machinery of emotion" rather than the circumstances leading to emotion. I expect the elucidation of emotions to tell us how feelings come about.

Emotions Precede Feelings

In discussing the precedence of emotion over feeling let me begin by calling attention to something Shakespeare left ambiguous in his lines for Richard. It has to do with the word shadow and with the possibility that while emotion and feeling are distinct, the latter comes before the former. The external laments are a shadow of the unseen grief, says Richard, some sort of mirror reflection of the principal object—the feeling of grief—just as Richard's face in the mirror is a reflection of the play's principal object, Richard. This ambiguity resonates well with one's untutored intuitions. We tend to believe that the hidden is the source of the expressed. Besides, we know that as far as the mind is concerned, feeling is what really counts. "There lies the substance," says Richard, speaking of his hidden grief, and we agree. We suffer or delight from actual feelings. In the narrow sense, emotions are externalities. But "principal" does not mean "first" and does not mean "causative." The centrality of feeling obscures the matter of how feelings arise and favors the view that somehow feelings occur first and are expressed subsequently in emotions. That view is incorrect, and it is to blame, at least in part, for the delay in finding a plausible neurobiological account for feelings.

It turns out that it is feelings that are mostly shadows of the external manner of emotions. Here is what Richard should have said, in effect, with due apologies to Shakespeare: "Oh, how this external manner of laments casts an intolerable and unseen shadow of grief in the silence of my tortured soul." (Which reminds me of James Joyce when he says in *Ulysses*, "Shakespeare is the happy hunting ground of all minds that have lost their balance."[4])

It is legitimate to ask at this point why emotions precede feelings. My answer is simple: We have emotions first and feelings after because evolution came up with emotions first and feelings later. Emotions are built from simple reactions that easily promote the survival of an organism and thus could easily prevail in evolution.

In brief, those whom the gods wanted to save they first made smart, or so it would seem. Long before living beings had anything like a creative intelligence, even before they had brains, it is as if nature decided that life was both very precious and very precarious. We know that nature does not operate by design and does not decide in the way artists and engineers do, but this image gets the point across. All living organisms from the humble amoeba to the human are born with devices designed to solve *automatically*, no proper reasoning required, the basic problems of life. Those problems are: finding sources of energy; incorporating and transforming energy; maintaining a chemical balance of the interior compatible with the life process; maintaining the organism's structure by repairing its wear and tear; and fending off external agents of disease and physical injury. The single word homeostasis is convenient shorthand for the ensemble of regulations and the resulting state of regulated life.[5]

In the course of evolution the innate and automated equipment of life governance—the homeostasis machine—became quite sophisticated. At the bottom of the organization of homeostasis we find simple responses such as *approaching* or *withdrawing* of an entire organism relative to some object; or increases in activity (*arousal*) or decreases in activity (*calm* or *quiescence*). Higher up in the organization we find *competitive* or *cooperative* responses.[6] We can picture the homeostasis machine as a large multibranched tree of phenomena charged with the automated regulation of life. In multicellular organisms, working our way from the ground up, here is what we will find in the tree.

In the lowest branches

- The process of metabolism. This includes chemical and mechanical components (e.g., endocrine/hormonal secretions; muscular contractions related to digestion, and so forth) aimed at maintaining the balance of internal chemistries. These reactions govern, for example, heart rate and blood pressure (which help the proper distribution of blood flow in the body); adjustments of acidity and alkalinity in the internal milieu (the fluids in the bloodstream and in the spaces between cells); and the storage and deployment of proteins, lipids, and carbohydrates required to supply the organism with energy (necessary for motion, manufacture of chemical enzymes, and maintenance and renewal of its structure).
- Basic reflexes. This includes the startle reflex, which organisms deploy in reaction to a noise or touch or as the tropisms or taxes that guide organisms away from extreme heat or extreme cold, away from dark and into light.
- The immune system. It is prepared to ward off viruses, bacteria, parasites, and toxic chemical molecules invading from outside the organism. Curiously, it also is prepared to deal with chemical molecules normally contained in healthy cells in the body that can become dangerous to the organism when released from dying cells into the internal milieu (e.g., breakdown of hyaluron; glutamate). In brief, the immune system is a first line of defense of the organism when its integrity is menaced from outside or from within.

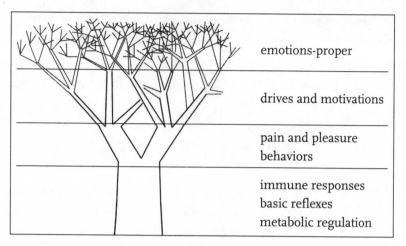

Figure 2.1: *Levels of automated homeostatic regulation, from simple to complex.*

In the middle-level branches

- Behaviors normally associated with the notion of pleasure (and reward) or pain (and punishment). These include reactions of approach or withdrawal of the whole organism relative to a specific object or situation. In humans, who can both feel and report what is felt, such reactions are described as painful or pleasurable, rewarding or punishing. For example, when there is malfunction and impending damage to tissues in the body—as happens in a local burn or infection—cells in the affected region emit chemical signals that are called nociceptive (this means "indicative of pain"). In response, the organism automatically reacts with *pain behaviors* or *sickness behaviors*. These are packages of actions, clearly visible or subtle, with which nature automatically counters the insult. Such actions include withdrawal of the whole body or a part thereof from the source of trouble if that source is external and identifiable; protection of the affected body part (holding a hand that has been wounded; hugging the chest or abdomen); and facial expressions of alarm and suffering. There also is a host of responses invisible to the naked eye and organized by the immune system. Those include increasing certain classes of white blood cells, dispatching those cells to the body areas in danger, and producing chemicals such as cytokines that help solve the problem the body is facing (fight off an invading microbe, repair damaged tissue). The ensemble of these actions and the chemical signals involved in their production form the basis for what we experience as *pain*.

In the same way the brain reacts to a problem in the body, it also reacts to the good function of that body. When the body operates smoothly, without hitch and with ease of transformation and utilization of energy, it behaves with a particular style. The approach to others is facilitated. There is relaxation and opening of the body frame, facial expressions of confidence and well-being, and production of certain classes of chemicals, such as endorphins, which are as invisible to the naked eye as some of the reactions in pain and sickness behaviors. The ensemble of these actions and the chemical signals associated with them form the basis for the experience of *pleasure*.

Pain or pleasure are prompted by many causes— glitches in some body function, optimal operation of metabolic regulation, or from external events that damage the organism or protect it. But the *experience* of pain or pleasure is *not the cause of the pain or pleasure behaviors*, and is by no means necessary for those behaviors to occur. As we will see in the next section, very simple creatures can carry out some of these emotive behaviors even if the likelihood of feeling those behaviors is low or nil.

In the next level up

- A number of *drives and motivations*. Major examples include hunger, thirst, curiosity and exploration, play and sex. Spinoza lumped them together under a very apt word, *appetites*, and with great refinement used another word, *desires*, for the situation in which conscious individuals become cognizant of those *appetites*. The word appetite designates the behavioral state of an organism engaged by a particular drive; the word desire refers to the conscious feelings of having an appetite and the eventual consummation or thwarting of the appetite. This Spinozian distinction is a nice counterpart for the distinction between emotion and feeling with which we started this chapter. Obviously humans have both appetites and desires just as seamlessly connected as emotions and feelings.

Near the top but not quite

- Emotions-proper. This is where we find the crown jewel of automated life regulation: emotions in the narrow sense of the term—from joy and sorrow and fear, to pride and shame and sympathy. And in case you wonder what we find at the very top, the answer is simple: feelings, which we will address in the next chapter.

The genome makes certain that all of these devices are active at birth, or shortly thereafter, with little or no dependence on learning, although as life continues learning will play an important role in determining *when* the devices are deployed. The more complex the reaction, the more this holds true. The package of reactions that constitutes crying and sobbing is ready and active at birth; what we cry *for*, across a lifetime, changes with our experience. All of these reactions are automatic and largely stereotyped, and are engaged under specific circumstances. (Learning, however, can modulate the execution of the stereotyped pattern. Our laughter or crying *plays* differently in different circumstances, just as the musical notes that constitute a movement of a sonata can be played in very different ways.) All of these reactions are aimed, in one way or another, directly or indirectly, at regulating the life process and promoting survival. Pleasure and pain behaviors, drives and motivations, and emotions-proper are sometimes referred to as emotions in the broad sense, which is both understandable and reasonable given their shared form and regulatory goal.[7]

Not content with the blessings of mere survival, nature seems to have had a nice afterthought: The innate equipment of life regulation does not aim for a neither-here-nor-there neutral state midway between life and death. Rather, the goal of the homeostasis endeavor is to provide a better than neutral life state, what we as thinking and affluent creatures identify as wellness and *well-being*.

The entire collection of homeostatic processes governs life moment by moment in every cell of our bodies. This governance is achieved by means of a simple arrangement: First, something changes in the environment of an individual organism, internally or externally. Second, the changes have the potential to alter the course of the life of the organism (they can constitute a threat to its integrity, or an opportunity for its improvement). Third, the organism detects the change and acts accordingly, in a manner designed to create the most beneficial situation for its own self-preservation and efficient functioning. All reactions operate under this arrangement and are thus a means to *appraise* the internal and external circumstances of an organism and act accordingly.

They detect trouble or detect opportunity and solve, by means of action, the problem of getting rid of the trouble or reaching out for the opportunity. Later, we shall see that even in "emotions-proper"—emotions such as sadness, or love, or guilt—the arrangement remains, except that the complexity of the appraisal and response are far greater than with the simple reactions from which such emotions were pieced together in evolution.

It is apparent that the continuous attempt at achieving a state of positively regulated life is a deep and defining part of our existence—the first reality of our existence as Spinoza intuited when he described the relentless endeavor (*conatus*) of each being to preserve itself. Striving, endeavor, and tendency are three words that come close to rendering the Latin term *conatus,* as used by Spinoza in Propositions 6, 7, and 8 of the *Ethics,* Part III. In Spinoza's own words: "Each thing, as far as it can by its own power, strives to persevere in its being" and "The striving by which each thing strives to persevere in its being is nothing but the actual essence of the thing." Interpreted with the advantages of current hindsight, Spinoza's notion implies that the living organism is constructed so as to maintain the coherence of its structures and functions against numerous life-threatening odds.

The conatus subsumes both the impetus for self-preservation in the face of danger and opportunities and the myriad actions of self-preservation that hold the parts of a body together. In spite of the transformations the body must undergo as it develops, renews its constituent parts, and ages, the conatus continues to form the *same* individual and respect the *same* structural design.

What is Spinoza's *conatus* in current biological terms? It is the aggregate of dispositions laid down in brain circuitry that, once engaged by internal or environmental conditions, seeks both survival and well-being. In the next chapter, we shall see how the large compass of activities of the *conatus* is conveyed to the brain,

chemically and neurally. This is accomplished by chemical molecules transported in the bloodstream, as well as by electrochemical signals transmitted along nerve pathways. Numerous aspects
of the life process can be so signaled to the brain and represented
there in numerous maps made of circuits of nerve cells located in
specific brain sites. By that point we have reached the treetops of
life regulation, the level at which feelings begin to coalesce.

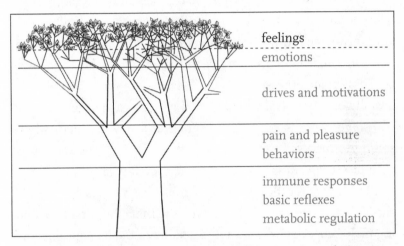

feelings

emotions

drives and motivations

pain and pleasure
behaviors

immune responses
basic reflexes
metabolic regulation

*Figure 2.2: Feelings support yet another level of homeostatic regulation.
Feelings are a mental expression of all other levels of homeostatic regulation.*

A Nesting Principle

When we survey the list of regulatory reactions that ensure our
homeostasis we glean a curious construction plan. It consists of
having parts of simpler reactions incorporated as components of
more elaborate ones, a nesting of the simple within the complex.
Some of the machinery of the immune system and of metabolic
regulation is incorporated in the machinery of pain and pleasure
behaviors. *Some* of the latter is incorporated in the machinery of
drives and motivations (most of which revolve around metabolic
corrections and all of which involve pain or pleasure). *Some* of the

machinery from all the prior levels—reflexes, immune responses, metabolic balancing, pain or pleasure behaviors, drives—is incorporated in the machinery of the emotions-proper. As we shall see, the different tiers of emotions-proper are assembled on the very same principle. The ensemble does not look exactly like a neat Russian doll because the bigger part is not merely an enlargement of the smaller part nested in it. Nature is never that tidy. But the "nesting" principle holds. Each of the different regulatory reactions we have been considering is not a radically different process, built from scratch for a specific purpose. Rather, each reaction consists of tinkered rearrangements of bits and parts of the simpler processes below. They are all aimed at the same overall goal—survival with well-being—but each of the tinkered rearrangements is secondarily aimed at a new problem whose solution is necessary for survival with well-being. The solution of each new problem is required for the overall goal to be achieved.

The image for the ensemble of these reactions is not that of a simple linear hierarchy. That is why the metaphor of a tall building with many floors only captures some of the biological reality. The image of the great chain of being is not good either. A better image is that of a tall, messy tree with progressively higher and more elaborate branches coming off the main trunks and thus maintaining a two-way communication with their roots. The history of evolution is written all over that tree.

More on the Emotion-Related Reactions: From Simple Homeostatic Regulation to Emotions-Proper

Some of the regulatory reactions we have been considering respond to an object or situation in the environment—a potentially dangerous situation; or an opportunity for feeding or mating. But some of the reactions respond to an object or situation *within* the organism. This can be a drop in the amount of available nutrients for the production of energy, causing the appetitive behaviors known as hunger and including the search for food. Or it could be a hormonal change that prompts the searching for a mate, or a wound that causes the reactions we call pain. The range of reactions encompasses not only highly visible emotions such as fear or anger, but also drives, motivations, and behaviors associated with pain or pleasure. They all occur within an organism, a body limited by a boundary, within which life ticks away. All of the reactions, directly or indirectly, exhibit an apparent aim: making the internal economy of life run smoothly. The amount of certain chemical molecules must be maintained within certain ranges, not higher and not lower, because outside those ranges life is in peril. Temperature also must be maintained within narrow parameters. Sources of energy must be procured—and curiosity and exploration strategies help locate those sources. Once found, those sources of energy must be incorporated—literally, placed inside the body—and modified for immediate consumption or storage; waste products resulting from all the modifications must be eliminated; and repair of the tissue wear and tear must be carried out so that the integrity of the organism is maintained.

Even the emotions-proper—disgust, fear, happiness, sadness, sympathy, and shame—aim directly at life regulation by staving off dangers or helping the organism take advantage of an opportunity, or indirectly by facilitating social relations. I am not suggesting every time we engage an emotion we are promoting survival and well-being. Not all emotions are alike in their potential to promote survival and well-being, and both the context in which an emotion is engaged and the intensity of the emotion are important factors in the potential value of an emotion on a specific occasion. But the fact that the deployment of some emotions in current human circumstances may be maladaptive does not deny their evolutionary role in advantageous life regulation.

Anger is mostly counterproductive in modern societies, and so is sadness. Phobias are a major hindrance. And yet think of how many lives have been saved by fear or anger in the right circumstances. These reactions are likely to have prevailed in evolution because they automatically supported survival. They still do, and that is probably why they remain part and parcel of the daily existence of human as well as nonhuman species.

On a practical note, understanding the biology of emotions and the fact that the value of each emotion differs so much in our current human environment, offers considerable opportunities for understanding human behavior. We can learn, for example, that some emotions are terrible advisors and consider how we can either suppress them or reduce the consequences of their advice. I am thinking, for example, that reactions that lead to racial and cultural prejudices are based in part on the automatic deployment of social emotions evolutionarily meant to detect *difference* in others because difference may signal risk or danger, and promote withdrawal or aggression. That sort of reaction probably achieved useful goals in a tribal society but is no longer useful, let alone appropriate, to ours. We can be wise to the fact that our brain still carries the machinery to react in the way it did in a very different context ages ago. And we can learn to disregard such reactions and persuade others to do the same.

The Emotions of Simple Organisms

There is abundant evidence of "emotional" reactions in simple organisms. Think of a lone paramecium, a simple unicellular organism, all body, no brain, no mind, swimming speedily away from a possible danger in a certain sector of its bath—maybe a poking needle, or too many vibrations, or too much heat, or too little. Or the paramecium may be swimming speedily along a chemical gradient of nutrients toward the sector of the bath where

it can have lunch. This simple organism is designed to detect certain signs of danger—steep variations in temperature, excessive
vibrations, or the contact of a piercing object that might rupture
its membrane—and react by proceeding to a safer, more temper-
ate, quieter place. Likewise, it will swim in the trail toward greener
water pastures after detecting the presence of chemical molecules
it needs for energy supply and chemical balance. The events I am
describing in a brainless creature already contain the essence of
the process of emotion that we humans have—detection of the
presence of an object or event that recommends avoidance and
evasion or endorsement and approach. The ability to react in this
manner was not taught—there is not much pedagogy going on in
paramecium school. It is contained in the apparently simple and
yet so complicated gene-given machinery inside the unbrained
paramecium. This shows that nature has long been concerned
with providing living organisms with the means to regulate and
maintain their lives automatically, no questions asked, no thoughts
needed.

Having a brain, even a modest brain, is helpful for survival, of
course, and indispensable if the environment is more challenging
than the paramecium's. Think of a tiny fly—a small creature with
a small nervous system but no spine. You can make the fly quite
angry if you swat it repeatedly and unsuccessfully. It will buzz
around you in daredevil supersonic dives and avoid the fatal swat.
But you also can make the fly happy if you feed it sugar. You can
see how its movements slow down and round themselves in response to comfort food. And you can make the fly giddily happy if
you give it alcohol. I am not inventing: The experiment has been
carried out on a fly species known as *Drosophila Melanogaster*.[8]
After exposure to ethanol vapor the flies are as uncoordinated as
we would be, given a comparable dose. They walk with the abandon of contented inebriation, and fall down an experimental tube

like drunks staggering to a lamppost. Flies have emotions, although I am not suggesting that they *feel* emotions, let alone that
they would reflect on such feelings. And if anyone is skeptical
about the sophistication of the life-regulation mechanisms in
such small creatures, consider the sleep mechanisms of the fly described by Ralph Greenspan and his colleagues.[9] Tiny *Drosophila*
has the equivalent of our day-night cycles, periods of intense activity and restorative sleep, and even the sort of response to sleep
deprivation that we show when we are jet-lagged. They need more
sleep, as do we.

Or think of the marine snail *Aplysia Californica*—again no
spine, little brain, and much sloth. Touch it in the gill and it will
fold into itself, increase its blood pressure, and jump up its heart
rate. The snail produces a number of concerted reactions that,
transposed to you or me, probably would be recognized as important components of the emotion fear. Emotion? Yes. Feeling?
Probably not.[10]

None of these organisms produce these reactions as a result
of deliberation. Nor do they *construct* the reaction either, bit by bit,
with some original flair for each instance in which the reaction
is displayed. The organisms react reflexively, automatically, in a
stereotypical fashion. Like the distracted shopper selecting from
a ready-to-wear display, they "select" ready-to-use responses and
move on. It would be incorrect to call these reactions reflexes because classical reflexes are simple responses, whereas these reactions are complex packages of responses. The multiplicity of
components and the coordination of the components distinguish
emotion-related reactions from reflexes. Better to say that they are
collections of reflex responses, some quite elaborate and all quite
well coordinated. They allow an organism to respond to certain
problems with an effective solution.

The Emotions-Proper

There is a venerable tradition of classifying emotions in varied categories. Although the classifications and labels are manifestly inadequate, there is no alternative at this point given the provisional stage of our knowledge. As knowledge accrues, the labels and the classifications are likely to change. In the meantime, we must remember that the borders between categories are porous. For the time being, I find it helpful to classify the emotions-proper in three tiers: background emotions, primary emotions, and social emotions.

As the term suggests, background emotions are not especially prominent in one's behavior, although they are remarkably important. You may never have paid much attention to it, but you probably are a good reader of background emotions if you accurately detect energy or enthusiasm in someone you have just met; or if you are capable of diagnosing subtle malaise or excitement, edginess or tranquillity, in your friends and colleagues. If you are really good, you can do the diagnostic job without a single word being uttered by your victim. You assess the contour of movements in the limbs and the entire body. How strong? How precise? How ample? How frequent? You observe facial expressions. If words do get uttered you do not just listen to the words and picture their dictionary meanings, you listen to the music in the voice, to the prosody.

Background emotions can be distinguished from moods, which refer to the sustaining of a given emotion over long periods of time, measured over many hours or days, such as when "Peter has been in a foul mood." Mood also can be applied to the frequently repeated engagement of the same emotion, such as when Jane, who is such a steady girl, "has been flying off the handle for no reason."

When I developed this notion,[11] I began seeing background emotions as the consequence of deploying certain combinations of the simpler regulatory reactions (e.g., basic homeostatic processes, pain and pleasure behaviors, and appetites), according to the nesting principle noted earlier. Background emotions are composite expressions of those regulatory actions as they unfold and intersect moment by moment in our lives. I imagine background emotions as the largely unpredictable result of several concurrent regulatory processes engaged within the vast playground that our organisms resemble. These include metabolic adjustments associated with whatever internal need is arising or has just been satisfied; and with whatever external situation is now being appraised and handled by other emotions, appetites, or intellectual calculation. The ever-changing result of this cauldron of interactions is our "state of being," good, bad, or somewhere in-between. When asked "how we feel," we consult this "state of being" and answer accordingly.

It is appropriate to ask if there are any regulatory reactions that do *not* contribute to background emotions; or which regulatory reactions are most frequently encountered in the makeup of background emotions such as discouragement or enthusiasm; or how do temperament and state of health interact with background emotions. The simple answer is that we do not know yet; the necessary investigations have not been done.

The *primary* (or basic) *emotions* are easier to define because there is an established tradition of lumping certain prominent emotions in this group. The frequent listing includes fear, anger, disgust, surprise, sadness, and happiness—the emotions that first come to mind whenever the term emotion is invoked. There are good reasons for this centrality. These emotions are easily identifiable in human beings across several cultures and in non-human species as well.[12] The circumstances that cause the emo-

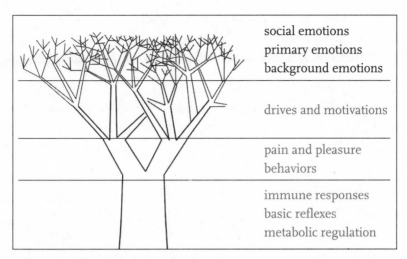

social emotions
primary emotions
background emotions

drives and motivations

pain and pleasure
behaviors

immune responses
basic reflexes
metabolic regulation

Figure 2.3: There are at least three kinds of emotion-proper: background emotions, primary emotions, and social emotions. The nesting principle applies here, too. For example, social emotions incorporate responses that are part of primary and background emotions.

tions and pattern of behaviors that define the emotions also are quite consistent across cultures and species. Not surprisingly, most of what we know about the neurobiology of emotion comes from studying the primary emotions.[13] Fear leads the way, as Alfred Hitchcock would have no doubt predicted, but notable strides are being made regarding disgust,[14] sadness and happiness.[15]

The *social emotions* include sympathy, embarrassment, shame, guilt, pride, jealousy, envy, gratitude, admiration, indignation, and contempt. The nesting principle applies to social emotions as well. A whole retinue of regulatory reactions along with elements present in primary emotions can be identified as subcomponents of social emotions in varied combinations. The nested incorporation of components from lower tiers is apparent. Think of how the social emotion "contempt" borrows the facial expressions of "disgust," a primary emotion that evolved in association with the automatic and beneficial rejection of potentially toxic foods. Even the words we use to describe situations of contempt, and moral outrage—we profess to be *disgusted*—revolve around the nesting. Pain and pleasure ingredients also are evident under the surface of social emotions, albeit subtler than in the primary emotions.

We are just beginning to understand how the brain triggers and executes the social emotions. Because the term "social" inevitably conjures up the notion of human society and of culture, it is important to note that social emotions are by no means confined to humans. Look around and you will find examples of social emotions in chimpanzees, baboons, and plain monkeys; in dolphins and lions; in wolves; and, of course, in your dog and cat. The examples abound—the proud ambulations of a dominant monkey; the literally regal deportment of a dominant great ape or wolf that commands the respect of the group; the humiliated behavior of the animal that does not dominate and must yield space and precedence at mealtimes; the sympathy an elephant shows toward another that is injured and ailing; or the embarrassment the dog shows after doing what he should not.[16]

Since none of these animals is likely to have been taught to emote, it appears that the disposition to exhibit a social emotion is ingrained deep in the organism's brain, ready to be deployed when the appropriate situation manages to trigger it. There is no doubt that the general brain arrangement that permits such sophisticated behaviors in the absence of language and instruments of culture is a gift of the genome of certain species. It is part of the roster of their largely innate and automated life-regulation devices, no less so than the others we have just discussed.

Does this mean these emotions are innate in the strict sense of the term and ready to be deployed immediately after birth in the same manner that metabolic regulation clearly is, after our first breath? The answer is likely to be different for different emotions. In some instances, emotional responses may be strictly innate; in others they may require minimal help from an appropriate exposure to the environment. Robert Hinde's work on fear is perhaps a good pointer as to what may happen in the social emotions. Hinde showed that the monkey's innate fear of snakes requires an exposure not just to a snake but to the mother's expression of fear of the snake. Once is enough for the behavior to kick into gear, but without that "once" the "innate" behavior is not engaged.[17] Something of this sort applies to the social emotions. An example is the establishment of patterns of dominance and submission in very young primates during play.

It remains difficult to accept, for anyone raised on the conviction that social behaviors are the necessary products of education, that simple animal species not known for their culture can exhibit intelligent social behaviors. But they do, and once again, they do not require that much brain to dazzle us. The modest worms C. elegans have exactly 302 neurons and about 5,000 interneuron connections. (For the sake of comparison, humans have several billion neurons and several trillion connections.) When these sexy little beasts (they are hermaphrodites!) are up and about in an environment with enough food and without stressors, they keep to themselves and feed in isolation. But if food is scarce or if a pestilent odor is present in the environment—by which I mean a threat if you lead a worm's existence and connect with the world through your nose—the worms congregate in single regions and feed together in groups. Just in case.[18] A number of curious social concepts are foreshadowed in this necessarily embryonic and yet

far-reaching behavior: safety in numbers, strength through cooperation, belt-tightening, altruism, and the original labor union.
Did you ever think humans invented such behavioral solutions?
Just consider the honeybee, small and very social in its hive society. A honeybee has 95,000 neurons. Now, that's a brain.

It is highly probable that the availability of such social emotions has played a role in the development of complex cultural
mechanisms of social regulation (see chapter 4). It also is apparent that some social emotional reactions are elicited in human
social situations without the stimulus for the reaction being immediately apparent to the reactor and to observers. Displays of
social dominance and dependence are an example—think of all
the strange antics of human behavior in sports, politics, and the
workplace. One of the many reasons why some people become
leaders and others followers, why some command respect and
others cower, has little to do with knowledge or skills and a lot to
do with how certain physical traits and the manner of a given individual promote certain emotional responses in others. To observers of such responses and to the individuals exhibiting them,
some of the displays appear unmotivated because they have their
origin in the innate, nonconscious apparatus of social emotion
and self-preservation. We should credit Darwin for leading us to
the evolutionary trail of these phenomena.

These are not the only emotional reactions of mysterious origin. There is another class of reactions with a nonconscious origin shaped by learning during one's individual development. I am
referring to the affinities and detestations we acquire discreetly in
the course of a lifetime of perceiving and emoting in relation to
people, groups, objects, activities, and places to which Freud
called our attention. Curiously, these two sets of nondeliberate,
nonconscious reactions—those innate and those learned—may
well be interrelated in the bottomless pit of our unconscious. One

is tempted to say that their possible nonconscious interplay signals the intersection of two intellectual legacies, that of Darwin and that of Freud, two thinkers who dedicated their work to studying the diverse influences of the innate and the acquired from below stairs.[19]

From chemical homeostatic processes to emotions-proper, life-regulation phenomena, without exception, have to do, directly or indirectly, with the integrity and health of the organism. Without exception, all of these phenomena are related to adaptive adjustments in body state and eventually lead to the changes in the brain mapping of body states, which form the basis for feelings. The nesting of the simple within the complex ensures that the regulatory purpose remains present in the higher echelons of the chain. While the purpose remains constant, complexity varies. Emotions-proper are certainly more complex than reflexes; and the triggering stimuli and target of the responses varies as well. The precise situations that initiate the process and their specific aim differ.

Hunger and thirst, for example, are simple appetites. The causative object is usually internal—a diminution in the availability of something vital for survival, namely, energy from food and water. But the ensuing behaviors are aimed at the environment and involve the search for the missing something, a search that involves exploratory motion of the surroundings and sensory detection of the thing being searched. This is not that different from what happens in emotions-proper, say, fear or anger. There, too, a competent object triggers the routine of adaptive behaviors. But the competent objects for fear and anger are almost always external (even when they are conjured up from memory and imagination in our brains they tend to stand for external objects), and

are quite varied in design (many kinds of physical stimulus, evolutionarily set or associatively learned, can cause fear). The most
frequent competent triggers for hunger or thirst tend to be internal (although we can become hungry or thirsty from watching
one more French movie in which the characters eat and drink
and are merry). Also some drives, at least in relation to nonhumans, are periodic and limited to seasons and physiological
cycles, e.g., sex, while emotions occur anytime and can be sustained over time.

We also discover curious interactions across classes of regulatory reactions. Emotions-proper influence appetites, and vice
versa. For example, the emotion fear inhibits hunger and sexual
drives, and so do sadness and disgust. On the contrary, happiness promotes both hunger and sexual drives. The satisfaction of
drives—hunger, thirst, and sex for example—causes happiness;
but thwarting the satisfaction of those drives can cause anger, or
despair, or sadness. Also, as noted earlier, the composite of the
daily unfolding of adaptive reactions, e.g., homeostatic adjustments and drives, constitutes the ongoing background emotions
and helps define mood over extended periods of time. Nonetheless, when you consider these different levels of regulatory reaction at some distance, one is struck by their remarkable formal
similarity.[20]

To the best of our knowledge, most of the living creatures
equipped to emote for the sake of their lives have no more brain
equipment to feel those emotions than they do to think of having
such emotions in the first place. They detect the presence of certain stimuli in the environment and respond to them with an
emotion. All they require is a simple perceptual apparatus—a fil-
ter to detect the emotionally competent stimulus and the capacity
to emote. Most living creatures act. They probably do not feel like
we do, let alone think like we do. This is a presumption, of course,

but it is justified by our idea of what it takes to feel as explained in the next chapter. The simpler creatures lack the brain structures necessary to portray in the form of sensory maps the transformations that occur in the body when emotive reactions take place and that result in feeling. They also lack the brain necessary to represent the anticipated simulation of such body transformations, which would constitute the basis for desire or anxiety.

It is apparent that the regulatory reactions discussed above are advantageous to the organism that exhibits them, and that the causes of those reactions—the objects or situations that trigger them—could be judged "good" or "bad" depending on their impact on survival or well-being. But it should be apparent that the paramecium or the fly or the squirrel do not know the good or evil qualities of these situations let alone consider acting for the "good" and against the "bad." Nor are we humans striving for goodness when we balance the pH in our internal milieu or react with happiness or fear to certain objects around us. Our organisms gravitate toward a "good" result of their own accord, sometimes directly as in a response of happiness, sometimes indirectly as in a response of fear that begins by avoiding "evil" and then results in "good." I am suggesting, and I will return to this point in chapter 4, that organisms can produce advantageous reactions that lead to good results without *deciding* to produce those reactions, and even without *feeling* the unfolding of those reactions. And it is apparent from the makeup of those reactions that, as they take place, the organism moves for a certain period toward states of greater or lesser physiological balance.

I offer qualified congratulations to us humans for two reasons. First, in comparable circumstances, these automated reactions create conditions in the human organism that, once mapped in the nervous system, can be represented as pleasurable or painful and eventually known as feelings. Let us say that this is the real source of human glory and human tragedy. Now for the second reason. We humans, conscious of the relation between certain objectives and certain emotions, can *willfully* strive to control our emotions, to some extent at least. We can decide which objects and situations we allow in our environment and on which objects and situations we lavish time and attention. We can, for example, decide not to watch commercial television, and advocate its eternal banishment from the households of intelligent citizens. By controlling our interaction with objects that cause emotions we are in effect exerting some control over the life process and leading the organism into greater or lesser harmony, as Spinoza would wish. We are in effect overriding the tyrannical automaticity and mindlessness of the emotional machinery. Curiously, humans long ago discovered this possibility without quite knowing the physiological basis for the strategies they use. This is what we do when we make choices regarding what we read or whom we befriend. This is what humans have done for centuries when they follow social and religious percepts that in effect modify the environment and our relation with it. This is what we try to do when we flirt with all the healthy living programs that make us exercise and diet.

It is not accurate to say that regulatory reactions including the emotions-proper are fatally and inevitably stereotyped. Some "low branch" reactions are and should be stereotyped—one does not want to interfere with nature's wisdom when it comes to regulating cardiac function or running away from danger. But the "high branch" reactions can be modified to some extent. We can control our exposure to the stimuli that bring on the reactions. We can learn over a lifetime to engage modulating "brakes" on those reactions. We can simply use sheer willpower and just say no. Sometimes.

A Hypothesis in the Form of a Definition

Taking the varied kinds of emotion in consideration, I can now offer a working hypothesis of emotion-proper in the form of a definition.

1. An emotion-proper, such as happiness, sadness, embarrassment, or sympathy, is a complex collection of chemical and neural responses forming a distinctive pattern.

2. The responses are produced by the normal brain when it detects an emotionally competent stimulus (an ECS), the object or event whose presence, actual or in mental recall, triggers the emotion. The responses are automatic.

3. The brain is prepared by evolution to respond to certain ECSs with specific repertoires of action. However, the list of ECSs is not confined to those prescribed by evolution. It includes many others learned in a lifetime of experience.

4. The immediate result of these responses is a temporary change in the state of the body proper, and in the state of the brain structures that map the body and support thinking.

5. The ultimate result of the responses, directly or indirectly, is the placement of the organism in circumstances conducive to survival and well-being.[21]

The classic components of an emotional reaction are encompassed by this definition, although the separation of the phases of the process and the weight accorded to those phases may appear unconventional. The process begins with an appraisal-evaluation phase, starting with the detection of an emotionally competent stimulus. My inquiry is focused on what happens after the stimulus is detected in the mind's process—the tail end of the appraisal phase. For obvious reasons, I also leave feelings, the next phase of the emotion-to-feeling cycle, out of the definition of emotion itself.

It might be argued, for the sake of functional purity, that the appraisal phase should be left out as well—appraisal being the process leading to emotion rather than emotion itself. But the radical excision of the appraisal phase would obscure rather than illuminate the real value of emotions: their largely intelligent connection between the emotionally competent stimulus and the set of reactions that can alter our body function and our thinking so profoundly. Leaving out appraisal also would render the biological description of the phenomena of emotion vulnerable to the caricature that emotions without an appraisal phase are meaningless events. It would be more difficult to see how beautiful and amazingly intelligent emotions can be, and how powerfully they can solve problems for us.[22]

The Brain Machinery of Emotion

Emotions provide a natural means for the brain and mind to evaluate the environment within and around the organism, and respond accordingly and adaptively. Indeed, in many circumstances, we actually evaluate consciously the objects that cause emotions, in the proper sense of the term "evaluate." We process not only the presence of an object but its relation to others and its connection to the past. In those circumstances the apparatus of emotions naturally evaluates, and the apparatus of the conscious mind thinkingly coevaluates. We even can modulate our emotional response. In effect, one of the key purposes of our educational development is to interpose a nonautomatic evaluative step between causative objects and emotional responses. We attempt by doing so to shape our natural emotional responses and bring them in line with the requirements of a given culture. All that is very true, but the point I wish to make here, however, is that in order for emotions to occur there is no *need* to analyze the causative object consciously let alone evaluate the situation in which it appears. Emotions can operate in different settings.

Even when the emotional reaction occurs without conscious knowledge of the emotionally competent stimulus the emotion signifies nonetheless the result of the organism's appraisal of the situation. Never mind that the appraisal is not made clearly known to the self. Somehow the notion of appraisal has been taken too literally to signify conscious evaluation, as if the splendid job of assessing a situation and responding to it automatically would be a minor biological achievement.

One of the main aspects of the history of human development pertains to how most objects that surround our brains become capable of triggering some form of emotion or another, weak or strong, good or bad, and can do so consciously or unconsciously. Some of these triggers are set by evolution, but some are not, instead becoming associated by our brains with emotionally competent objects by virtue of our individual experiences. Think of the house where once, as a child, you may have had an experience of intense fear. When you visit that house today you may feel uncomfortable without any cause for the discomfort other than the fact that, long ago, you had a powerful negative emotion in those same surroundings. It may even happen that in a different but somewhat similar house you experience the same discomfort, again for no reason other than you can detect the brain's record of a comparable object and situation.

There is nothing in your brain's basic makeup prepared to respond with displeasure to houses of a certain kind. But your life experience has made your brain associate such houses with the displeasure you once had. Never mind that the cause of the displeasure had nothing to do with the house itself. Call it guilt by association. The house is an innocent bystander. You have been *conditioned* to feel uncomfortable in certain houses, perhaps even to dislike certain houses without really knowing why. Or to feel well in certain houses, by precisely the same mechanism. Many of our perfectly normal and banal likes and dislikes arise this way. But note that phobias, which are neither normal nor banal, can be acquired by the same mechanism. At any rate, by the time we are old enough to write books, few if any objects in the world are emotionally neutral. The emotional distinction among objects is a distinction of grades: Some objects evoke weak, barely perceptible emotional reactions, some objects evoke strong emotional reactions, and there is every other grade in between. We even are beginning to uncover the molecular and cellular mechanisms necessary for emotional learning to occur.[23]

Complex organisms also learn to modulate the execution of emotions in harmony with the individual circumstances—and here the terms appraisal and evaluation are most apt. The emotional modulation devices can adjust the magnitude of emotional expression without an organism's conscious deliberation. One simple example: After being told the same amusing story for the second time you will smile or laugh quite differently depending on the social context of the moment—a diplomatic dinner, a casual hallway encounter, Thanksgiving dinner with close friends, and so on. If your parents have done a good job you will not need to *think* about the context. The adjustment is automatic. Some of the adjuster devices, however, do reflect a judgment on the part of the organism's self and may result in an attempt to modify or even suppress emotions. For a number of reasons that range from the honorable to the despicable, you may elect to conceal your disgust or mirth regarding some statement that a colleague or the person you are negotiating with just made. Conscious knowledge of the context and awareness of the future consequences of every aspect of your own behavior help you decide to suppress the natural expression of emotion. But try to avoid it as you get older. It is very energy consuming.

Emotionally competent objects can be actual or recalled from memory. We have seen how a nonconscious conditioned memory can lead to a current emotion. But memory can play the same trick out in the open. For example, the actual near-accident that frightened you years ago can be recalled from memory and cause you to be frightened anew. Whether actually present, as a freshly minted image, or as a reconstructed image recalled from memory, the kind of effect is the same. If the stimulus is emotionally competent an emotion ensues, and only the intensity varies. Actors of every sort of schooling rely on this so-called emotional memory for their trade. In some cases they let memory overtly lead them to emote. In other cases they let memory infiltrate their performance subtly, setting themselves up to behave in a certain way. Our ever-observant Spinoza did not leave this one alone either: *A man is as much affected pleasurably or painfully by the image of a thing past or future, as by the image of a thing present* [*Ethics*, Part III, Proposition 28].

Triggering and Executing Emotions

The appearance of an emotion depends on a complicated chain of events. Here is how I see it. The chain begins with the appearance of the emotionally competent stimulus. The stimulus, a certain object or situation actually present or recalled from memory, comes to mind. Think of the bear you came across on your trip to Alaska (this in homage to William James who wove his discussion of fear on the sighting of one such bear). Or think of a forthcoming meeting with someone you miss.

In neural terms, images related to the emotionally competent object must be represented in one or more of the brain's sensory

processing systems, such as the visual or auditory regions. Let us
call this the presentation stage of the process. Regardless of how
fleeting the presentation, signals related to the presence of that
stimulus are made available to a number of emotion-triggering
sites elsewhere in the brain. You can conceive of those sites as
locks that open only if the appropriate keys fit. The emotionally
competent stimuli are the keys, of course. Note that they select a
preexisting lock, rather than instruct the brain on how to create
one. The emotion-triggering sites subsequently activate a number
of emotion-execution sites elsewhere in the brain. The latter sites
are the immediate cause of the emotional state that occurs in
the body and in brain regions that support the emotion-feeling
process. Eventually, the process can reverberate and amplify itself, or shrivel away and close down. In the language of neuroanatomy and neurophysiology, this process begins when neural
signals of a certain configuration (that originate in visual cortices
that are holding neural patterns corresponding to the fast approach of a threatening object) are relayed in parallel along several
pathways to several brain structures. Some of the recipient structures, for example, the amygdala, will become active when they
"detect" a certain configuration—when the key fits the lock—and
initiate signals toward other brain regions, thus giving rise to a
cascade of events that will *become* an emotion.

These descriptions sound a lot like that of an antigen (e.g., a
virus) entering the bloodstream and leading to an immune response (consisting of a large number of antibodies capable of neu-
tralizing the antigen). And well they should because the processes
are formally similar. In the case of emotion the "antigen" is presented through the sensory system and the "antibody" is the emotional response. The "selection" is made at one of several brain
sites equipped to trigger an emotion. The conditions in which the
process occurs are comparable, the contour of the process is the
same, and the results just as beneficial. Nature is not that inven-

tive when it comes to successful solutions. Once it works, it tries it again and again. If only things would work as well for Hollywood producers, sequels would always make money.

Some of the brain regions now identified as emotion-triggering sites are the amygdala, deep in the temporal lobe; a part of the frontal lobe known as the ventromedial prefrontal cortex; and yet another frontal region in the supplementary motor area and cingulate. They are not the only triggering sites, but so far they are the best understood. These "triggering" sites are responsive to both natural stimuli, the electrochemical patterns that support the images in our minds, and to very unnatural stimuli, such as an electric current applied to the brain. But the sites should not be seen as rigid, delivering the same stereotyped performance time after time, because a number of influences can modulate their activity. Again, simple images in the mind as well as direct stimulation of brain structures can do the trick.

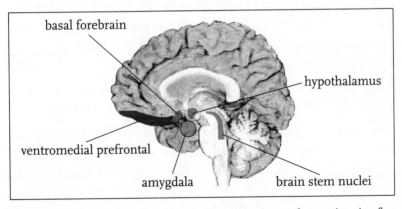

Figure 2.4: A minimalist view of the brain's triggering and execution sites for emotion. A large variety of emotions can be triggered when activity elsewhere in the brain induces activity in one of these sites, e.g., in parts of the amygdala or the ventromedial prefrontal cortex. None of these triggering sites produces an emotion by itself. For an emotion to occur the site must cause subsequent activity in other sites, e.g., in basal forebrain, hypothalamus, or nuclei of the brainstem. As with any other form of complex behavior, emotion results from the concerted participation of several sites within a brain system.

The study of the amygdala in animals has yielded important new information, most notably in the work of Joseph LeDoux, and modern brain imaging techniques have made studies of the human amygdala possible too, as exemplified by the studies of Ralph Adolphs and those of Raymond Dolan.[24] Those studies suggest that the amygdala is an important interface between visual and auditory emotionally competent stimuli and the triggering of emotions, in particular, though not exclusively, fear and anger. Neurological patients with damage to the amygdala cannot trigger those emotions and as a result do not have the corresponding feelings either. The locks for fear and anger seem to be missing, at least for visual and auditory triggers operating under regular circumstances. Recent studies also show that when recordings are made directly from single neurons in the human amygdala, a larger proportion of neurons are tuned to unpleasant stimuli than to pleasant.[25]

Curiously, the normal amygdala serves some of its triggering functions whether we are aware of the presence of an emotionally competent stimulus. Evidence for the amygdala's ability to detect emotionally competent stimuli nonconsciously first came from the work of Paul Whalen. When he showed such stimuli very rapidly to normal people who were entirely unaware of what they were seeing, brain scans revealed that the amygdala became active.[26] Recent work from Arnie Ohman and Raymond Dolan has shown that normal subjects can learn, covertly, that a certain stimulus but not another (e.g., a particular angry face but not another angry face) is associated with an unpleasant event. The covert representation of the face associated with the bad event prompts the activation of the *right* amygdala; but the covert representation of the other face does not.[27]

Emotionally competent stimuli are detected very fast, ahead of selective attention, as shown by an impressive finding: after le-

sions of the occipital lobe or parietal lobe cause a blind field of vision (or a field of vision in which stimuli are *not* detected due to
neglect), emotionally competent stimuli (e.g., angry or happy
faces) nevertheless "break through" the barrier of blindness or
neglect and are indeed detected.[28] The triggering emotional machinery captures these stimuli because they bypass the normal
processing channels—channels that might have led to cognitive
appraisal but simply could not do so because of blindness or
neglect. The value of this "bypass" biological arrangement is
apparent: whether one is paying attention, emotionally competent
stimuli *can be detected*. Subsequently, attention and proper thought
can be diverted to those stimuli.

Another important triggering site is in the frontal lobe, especially in the ventromedial prefrontal region. This region is tuned
to detecting the emotional significance of more complex stimuli,
for example objects and situations, natural as well as learned,
competent to trigger social emotions. The sympathy evoked by
witnessing someone else's accident, as well as the sadness evoked
by one's personal loss, require the mediation of this region. Many
of the stimuli that acquire their emotional significance in one's
life experience—as in the example of the house that becomes a
source of unpleasantness—trigger the respective emotions via
this region.

My colleagues Antoine Bechara, Hanna Damasio, and Daniel
Tranel and I have shown that damage to the frontal lobe alters the
ability to emote when the emotionally competent stimulus is social in nature, and when the appropriate response is a social emotion such as embarrassment, guilt, or despair. Impairments of
this sort compromise normal social behavior.[29]

In a recent series of studies from our group, Ralph Adolphs
has shown that neurons in the ventromedial prefrontal regions
respond rapidly and differently to the pleasant or unpleasant

emotional content of pictures. Single-cell recordings from the ventromedial prefrontal region of neurological patients being assessed for the surgical treatment of seizures reveal that numerous neurons in this region, and more so in the right frontal region than in the left, respond dramatically to pictures capable of inducing *unpleasant* emotions. They begin to react as early as 120 milliseconds after the stimulus is presented. First they suspend their spontaneous firing pattern; then, after a silent interval, they fire more intensely and more frequently. Fewer neurons respond to pictures capable of inducing *pleasant* emotions, and do so without the stop-and-go pattern noted for the unpleasantly tuned neurons.[30] The right-left brain asymmetry is more extreme than I would have predicted, but it is in keeping with a proposal made by Richard Davidson several years ago. Based on electroencephalographic studies conducted in normal individuals, Davidson suggested that the right frontal cortices were more associated with negative emotions than the left.

In order to create an emotional state, the activity in triggering sites must be propagated to execution sites by means of neural connections. The emotion-execution sites identified to date include the hypothalamus, the basal forebrain, and some nuclei in the brain stem tegmentum. The hypothalamus is the master executor of many chemical responses that are part and parcel of emotions. Directly or via the pituitary gland it releases into the bloodstream chemical molecules that alter the internal milieu, the function of viscera, and the function of the central nervous system itself. Oxytocin and vasopressin, both peptides, are examples of molecules released under the control of hypothalamic nuclei with the help of the posterior pituitary gland. A host of emotional behaviors (such as attachment and nurturing) depends on the timely availability of these hormones within the brain structures that command the execution of those behaviors. Likewise, the local brain availability of molecules such as dopamine and serotonin, which modulate neural activity, causes certain behaviors to occur. For example, the sort of behaviors experienced as rewarding and pleasurable appears to depend on the release of dopamine from one particular area (the ventrotegmental area in the brain stem), and its availability in yet another area (the nucleus accumbens in the basal forebrain). In short, the basal forebrain and hypothalamic nuclei, some nuclei in the brain stem tegmentum, and the brain stem nuclei that control the movement of the face, tongue, pharynx, and larynx are the ultimate executors of many behaviors, simple as well as complex, that define the emotions, from courting or fleeing to laughing and crying. The complex repertoires of actions we observe are the result of the exquisite coordination of the activities of those nuclei that contribute parts of the execution in a well-concerted order and concurrence. Jaak Panksepp has dedicated a lifetime of research to this execution process.[31]

In all emotions, multiple volleys of neural and chemical responses change the internal milieu, the viscera, and the musculoskeletal system for a certain period and in a particular pattern. Facial expressions, vocalizations, body postures, and specific patterns of behavior (running, freezing, courting, or parenting) are thus enacted. The body chemistries as well as viscera such as the heart and lungs help along. Emotion is all about transition and commotion, sometimes real bodily upheaval. In a parallel set of commands the brain structures that support image-production and attention change as well; as a result, some areas of the cerebral cortex appear to be less active, while others become especially so.

In the simplest of diagrams, here is how a visually presented threatening stimulus triggers the emotion fear and leads to its execution.

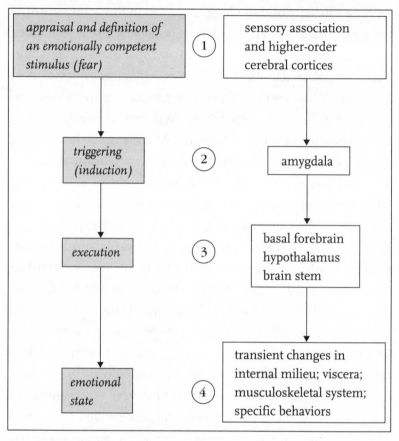

Figure 2.5: A diagram of the main stages for the triggering and execution of an emotion, using fear as an example. The shadow boxes on the left vertical column identify the stages of the process (1 to 3), from the appraisal and definition of the emotionally competent stimulus to the full-blown emotional state of fear (4). The boxes on the right vertical column identify the brain structures that are most necessary for each stage to unfold, and the physiological consequences of this chain of events (4).

For the purposes of providing a manageable description of the processes of emotion and feeling, I have simplified them to fit into a single chain of events beginning with a single stimulus and terminating with the establishment of the substrates of the feeling related to the stimulus. In reality, as might be expected, the process spreads laterally into parallel chains of events and amplifies itself. This is because the presence of the initial emotionally competent stimulus often leads to the recall of other related stimuli that are also emotionally competent. As time unfolds, the additional competent stimuli may sustain the triggering of the same emotion, trigger modifications of it, or even induce conflicting emotions. Relative to that initial stimulus, the continuation and intensity of the emotional state is thus at the mercy of the ongoing cognitive process. The contents of the mind either provide further triggers for the emotional reactions or remove those triggers, and the consequence is either the sustaining or even amplification of the emotion, or else its abatement.

The processing of emotions involves this dual track: the flowing of mental contents that bring along the triggers for the emotional responses, and the executed responses themselves, those that constitute emotions, which eventually lead to feelings. The chain that begins with the triggering of emotion and continues with the execution of emotion continues with the establishment of the substrates for feeling in the appropriate body-sensing brain regions.

Curiously, by the time the process reaches the stage of assembling feelings, we are back in the mental realm—back in the flow of thoughts where, in normal circumstances, the entire emotional detour began. Feelings are just as mental as the objects or events that trigger the emotions. What makes feelings distinctive as mental phenomena is their particular origin and content, the state of the organism's body, actual or as mapped in body-sensing brain regions.

Out of the Blue

Recently, a number of neurological studies have given us a closer look at the machinery that controls the execution of emotions. One of the most telling observations was made in a sixty-five-year-old woman undergoing treatment for Parkinson's disease. Nothing had suggested that in the course of attempting to relieve her symptoms we would be given a glimpse of how emotions come into being and of how they relate to feeling.

Parkinson's disease is a common neurological disorder that compromises the ability to move normally. Rather than causing paralysis, the condition causes rigidity of the muscles, tremors, and, perhaps most importantly, akinesia, a difficulty in initiating movements. Movements often are slow, a symptom known as bradykinesia. The disease used to be incurable, but for the past three decades it has been possible to alleviate the symptoms with the use of a medication containing levodopa, a chemical precursor of the neurotransmitter dopamine. Dopamine is missing in certain brain circuits of Parkinson's patients, much as insulin is missing in the bloodstream of patients with diabetes. (The neurons that produce dopamine in the pars compacta of the substantia nigra die away and dopamine is no longer made available at yet another brain region, the basal ganglia.) Unfortunately, the medications designed to increase dopamine in the brain circuits where it is missing do not help all patients. Also, in those that are helped, the medications may lose their effectiveness over time or cause other alterations of movement that are no less disabling than the disease. For this reason, several other modalities of treatment are being developed, one of which appears especially promising. It involves implanting tiny electrodes in the brain stem of Parkinson's patients so that the passage of a low-intensity, high-frequency electrical current can change the way in which some of the motor nuclei operate. The results usually are stunning. As the current passes, the symptoms vanish magically. The patients can move their hands with precision and walk so normally that a stranger might not be able to tell that something had previously been wrong.

The precise placement of the array of electrode contacts is a key to the success of the treatment. To achieve this, the surgeon uses a stereotaxic device (an apparatus that permits the localization of a brain structure in three-dimensional space) and carefully navigates the electrodes into the part of the brain stem known as the mesencephalon. There are two long, vertically oriented electrodes, one for the left side of the brain stem, another for the right, and each electrode has four contacts. The contacts are located about two millimeters from each other and each contact can be independently stimulated by the passage of an electrical current. By attempting stimulation at each contact site, it is possible to determine which contact produces the greatest degree of improvement without unwanted symptoms.

The intriguing story I am about to tell you involved a patient studied by my colleague Yves Agid and his team at the Salpêtrière Hospital in Paris. The patient was a sixty-five-year-old woman with a long history of parkinsonian symptoms that no longer responded to levodopa. She had no history of depression before or after the onset of the disease, and she had not even experienced mood changes, a common side effect of levodopa. She had no history of psychiatric disorder, personally or in her family.

Once the electrodes were in place, the procedure initially went the same way it had for nineteen other patients treated by the same group. The doctors found one electrode contact that greatly relieved the woman's symptoms. But the unexpected happened when the electric current passed through one of the four contact sites on the patient's left side, precisely two millimeters below the contact that improved her condition. The patient stopped her ongoing conversation quite abruptly, cast her eyes down and to her right side, then leaned slightly to the right and her emotional expression became one of sadness. After a few seconds she suddenly began to cry. Tears flowed and her entire demeanor was one of profound misery. Soon she was sobbing. As this display continued she began talking about how deeply sad she felt, how she had no energies left to go on living in this manner, how hopeless and exhausted she was. Asked about what was happening, her words were quite telling:

I'm falling down in my head, I no longer wish to live, to see anything, hear anything, feel anything...

I'm fed up with life, I've had enough... I don't want to live anymore, I'm disgusted with life...

Everything is useless... I feel worthless.

I'm scared in this world.

I want to hide in a corner... I'm crying over myself, of course... I'm hopeless, why am I bothering you?

The physician in charge of the treatment realized that this unusual event was due to the current and aborted the procedure. About ninety seconds after the current was interrupted the patient's behavior returned to normal. The sobbing stopped as abruptly as it had begun. The sadness vanished from the patient's face. The verbal reports of sadness also terminated. Very rapidly, she smiled, appeared relaxed, and for the next five minutes was quite playful, even jocular. What was that all about? she asked. She had felt awful but did not know why. What had provoked her uncontrollable despair? She was as puzzled as the observers were.

Yet the answer to her questions was clear enough. The electrical current had not passed into the general motor control structures as intended, but had flowed instead into one of the brain stem nuclei that control particular types of action. Those actions, as an ensemble, produce the emotion sadness. This repertoire included movements of the facial musculature; movements of the mouth, pharynx, larynx, and diaphragm, which are necessary for crying and sobbing; and the varied actions that result in the production and elimination of tears.

Remarkably, it appeared as if a switch had been turned on inside the brain in response to the switch that had been turned on outside of it. This entire repertoire of actions was engaged in a well-rehearsed instrumental concert, every step in its own time and place so that the effect appeared to manifest, for all intents and purposes, the presence of thoughts capable of causing sadness—the presence of emotionally competent stimuli. Except, of course, that no such thoughts had been present prior to the unexpected incident, nor was the patient even prone to having such thoughts spontaneously. Emotion-related thoughts only came *after* the emotion began.

Hamlet may wonder at the player's capability of conjuring up emotion in spite of having no personal cause for it. "Is it not monstrous that this player here, but in a fiction, in a dream of passion, could force his soul so to his own conceit, that from her working all his visage waned, tears in his eyes, distraction in his aspect, a broken voice, and his whole form suiting with forms to his own conceit?" The player has no personal cause whatever to be emotional—he is talking about the fate of a character called Hecuba, and, as Hamlet says, "What's Hecuba to him, or he to Hecuba." However, the player does begin by conjuring up some sad thoughts in his mind, which subsequently trigger the emotion and help him enact it with his artistry. Not so, in the strange case of this patient. There was no "conceit" prior to her emotion. There were no thoughts whatsoever to induce her behavior, no troubling ideas that came to her mind spontaneously, and no troubling ideas that she was asked to conjure up. The display of sadness, in all its spectacular complexity, came truly out of nowhere. No less importantly, sometime *after* the display of sadness was fully organized and in progress, the patient began to have a *feeling* of sadness. And, just as importantly, after she reported feeling sad she began having thoughts consonant with sadness—concern for her medical condition, fatigue, disappointment with her life, despair, and a wish to die.

The sequence of events in this patient reveals that the emotion sadness came first. The feeling of sadness followed, accompanied by thoughts of the type that usually can cause and then accompany the emotion sadness, thoughts that are characteristic of the states of mind we colloquially describe as "feeling sad." Once the stimulation ceased these manifestations waned and then vanished. The emotion disappeared and so did the feeling. The troubling thoughts were gone as well.

The importance of this rare neurological incident is apparent. In normal conditions the speed with which emotions arise and give way to feelings and related thoughts makes it difficult to analyze the proper sequence of phenomena. As thoughts normally causative of emotions appear in the mind, they cause emotions, which give rise to feelings, which conjure up other thoughts that are thematically related and likely to amplify the emotional state. The thoughts that are conjured up may even function as independent triggers for additional emotions and thus potentiate the ongoing affective state. More emotion gives rise to more feeling, and the cycle continues until distraction or reason put an end to it. By the time all these sets of phenomena are in full swing— the thoughts that can cause emotion; the behaviors of emotion; the mind phenomena we call feelings; and the thoughts that are consequent to feelings—it is difficult to tell by introspection what came first. This woman's case helps us see through the conflation. She had no thoughts causative of sadness or any feelings of sadness prior to having an emotion called sadness. The evidence speaks both to the relative autonomy of the neural triggering mechanism of emotion and to the dependence of feeling on emotion.

At this point one should ask: Why would this patient's brain evoke the kind of thoughts that normally cause sadness considering that the emotion and feeling were unmotivated by the appropriate stimuli? The answer has to do with the dependence of feeling on emotion and the intriguing ways of one's memory. When the emotion sadness is deployed, feelings of sadness instantly follow. In short order, the brain also brings forth the *kind* of thoughts that normally cause the emotion sadness *and* feelings of sadness. This is because associative learning has linked emotions with thoughts in a rich two-way network. Certain thoughts evoke certain emotions and vice versa. Cognitive and emotional levels of processing are continuously linked in this manner. This effect can be demonstrated experimentally as shown in a study by Paul Ekman and his colleagues. He asked subjects to move certain muscles of the face in a certain sequence, such that, unbeknownst to the subjects, the expression became one of happiness or sadness, or fear. The subjects did not know which expression was being portrayed on their faces. In their minds there was no thought capable of causing the portrayed emotion. And yet the subjects came to feel the feeling appropriate to the emotion displayed.[32] Without a doubt, parts of the emotion pattern came first. They were under the control of the experimenter and were not motivated by the subject. Some feeling followed thereafter. All of which conforms to the wisdom of Rodgers and Hammerstein. Remember that they have Anna (she who came to Siam to teach the King's children) telling her frightened self and her frightened son that whistling a happy tune will turn fear into confidence: "The results of this deception are very strange to tell. For when I fool the people I fear, I fool myself as well." Psychologically unmotivated and "acted" emotional expressions have the power to cause feeling. The expressions conjure up the feelings and the kinds of thoughts that have been learned as consonant with those emotional expressions.

From a subjective standpoint, the state of this patient after the activation of electrode "zero left" somewhat resembles those situations in which we find ourselves aware of moods and feelings, but unable to find the cause. How many times do we note, at a certain moment of a given day, that we are feeling especially well and filled with energy and hope, but don't know the reason; or, on the contrary, that we are feeling blue and edgy? In those instances, it is likely that troubling thoughts or hopeful thoughts are being processed outside of our field of consciousness. They are, nonetheless, capable of triggering the machinery of emotion and hence that of feeling. Sometimes we come to realize the origin of those affective states, sometimes we do not. For a good part of the twentieth century, many rushed to the psychoanalyst's couch to find out more about unconscious thoughts and about the equally unconscious conflicts that were giving rise to them. These days many people just accept that there are more unknown thoughts in the heaven and earth of our minds than Hamlet's friend Horatio could ever conceive in his philosophy. And when we cannot identify the emotion-causing thought, we are visited by unexplained emotions and feelings. Fortunately those emotions and feelings are less intense and less abrupt.

The group of physicians and investigators responsible for the patient's care further investigated her unusual case.[33] Stimulation at any of the other electrode contacts implanted in this same patient caused nothing unexpected, and as noted, this reaction did not occur in any other of the nineteen patients treated the same way. On two other occasions, and with the patient's appropriate consent, the doctors established the following facts. First, when they told the patient they were stimulating the problematic electrode contact, but actually were only clicking the switch for another electrode, no behavior whatsoever ensued. They observed nothing unusual and the patient reported nothing unusual. Second, when the problematic contact was switched on again, without warning, they reproduced the same set of events as in the original, unexpected observation. Electrode placement and electrode activation clearly were linked to the appearance of the phenomenon.

The investigators also carried out a functional imaging study (using positron-emission tomography) following stimulation of contact zero left. An important finding in the latter study was the marked activation of structures in the right parietal lobe, a region involved in the mapping of the body state and particularly of the mapping of the body in space. This activation probably related to the patient's consistent report during stimulation of marked changes in her body state, including the sensation of falling through a hole.

The scientific value of single-subject studies is always limited. The evidence usually is a starting point for new hypotheses and explorations rather than the endpoint of an investigation. Nonetheless, the evidence in this case is quite valuable. It supports the notion that the processes of emotion and feeling can be analyzed by component. It also reinforces a fundamental notion of cognitive neuroscience: Any complex mental function results from concerted contributions by *many* brain regions at varied levels of the central nervous system rather than from the work of a single brain region conceived in a phrenological manner.

The Brain Stem Switch

It is by no means clear which particular brain stem nucleus started the emotional reaction of this patient. The problematic contact appears to have been directly over the substantia nigra, but the current itself may have passed elsewhere in the vicinity. The brain stem is a very small region of the central nervous system and is jam-packed with nuclei and circuitry involved in different functions. Some of these nuclei are tiny and a minimal variation in the standard anatomy could have led to a significant rerouting of the current. But it is not in question that the event began in the mesencephalon and gradually recruited the nuclei required to produce several components of the emotion. It is even possible, judging from what has been gathered in animal experiments, that nuclei in the region known as the periaqueductal gray (PAG) were involved in the well-coordinated production of the emotion. We know, for example, that different columns of the PAG are involved in producing different kinds of fear reaction—the kind that ends up in fight-and-flight behaviors or, instead, in freezing behaviors. The PAG may be involved in sadness reactions as well. At any rate, within one of the emotion-related mesencephalic nuclei, a chain began that, quite rapidly, engaged extensive regions of the body—face, vocal apparatus, chest cavity, not to mention the chemical systems whose activities could not be observed directly. The changes led to a specific feeling state. Moreover, as the emotion sadness and the feelings of sadness unfolded, the patient recalled thoughts consonant with sadness. Instead of beginning in the cerebral cortex, the chain of events began in a subcortical region. But the effects were similar to those that would have been produced by thinking of a tragic event or witnessing it. Anyone who would have come on the scene at that point would not have been able to tell whether this was a perfectly natural emotion-feeling state, an emotion-feeling state created by the skills of a consummate actress, or an emotion-feeling state started by an electrical switch.

Out-of-the-Blue Laughter

Lest one would think that there is something unique about crying and sadness, I must add that a phenomenon equivalent to the one we have just analyzed can be produced for laughter, as shown in a study led by Itzhak Fried.[34] The circumstances also involved a patient undergoing electrical brain stimulation. The purpose was only slightly different: the mapping of cerebral cortex functions.

In order to help patients whose epileptic seizures do not respond to medications, it is possible to surgically remove the brain region that causes the seizures. In advance of surgery, however, the surgeon not only must localize with precision the area of brain that should be removed, but also must identify brain areas that cannot be removed because of their function, such as speech-related areas. This is achieved by stimulating the brain with electricity and observing the results.

In the particular case of patient A. K. when surgeons began stimulation in a region of the left frontal lobe known as the supplementary motor area (SMA), they noted that electrical stimulation at a number of closely located sites consistently and exclusively evoked laughter. The laughter was quite genuine, so much so that the observers described it as contagious. It came entirely out of the blue—the patient was not being shown or told anything funny, and was not entertaining any thought that might lead to laughter. And yet, there it was, entirely unmotivated but realistic laughter. Remarkably, and precisely as noted in the crying patient, laughter was followed "by a sensation of merriment or mirth" in spite of its unmotivated nature. Just as interestingly, the cause of the laughter was attributed to whichever object the patient was concentrating on at the time of the stimulation. For example, if the patient was being shown a picture of a horse, she would say, "The horse is funny." On occasion the investigators themselves were deemed to be an emotionally competent stimulus as when she concluded: "You guys are just so funny... standing around."

The laughter-producing brain patch was small, measuring about two centimeters by two centimeters. At nearby points, stimulation caused the well-known phenomena of either speech arrest or cessation of hand movements. However, such stimulations never caused laughter. Moreover, it should be noted that when this patient had seizures they never included laughter.

In the perspective of the framework described earlier, I believe that stimulation in the sites identified in this study leads to activity in the nuclei of the brain stem capable of producing the motor patterns of laughter. The precise brain stem nuclei and their sequence of activities have not been identified for either laughter or crying. Taken together these studies offer a glimpse of a multitiered neural mechanism for the production of emotions. After the processing of an emotionally competent stimulus, cortical sites initiate the actual emoting by triggering activity in other sites, largely subcortical, from where the execution of the emotion ultimately can be carried out. In the case of laughter, it appears that the initial triggering sites are in the medial and dorsal prefrontal region in regions such as the SMA and the anterior cingulate cortex. In the case of crying, the critical triggering sites are more likely to be in the medial and ventral prefrontal region. In both laughter and crying, the main execution sites are in brain stem nuclei. Incidentally, the evidence uncovered in the laughter study is in accordance with our own observations in patients with damage to the SMA and anterior cingulate. We have found that such patients have difficulty smiling a "natural" smile—a smile spontaneously induced by getting a joke—and that they are limited to the fake sort of "say cheese" smile.[35]

The studies discussed here testify to the separability of stages and mechanisms in the emotion and feeling process—appraisal/ evaluations leading to the isolation of an emotionally competent stimulus, triggering, execution, and subsequent feeling. The artificial electrical stimulus involved in the laughter study mimics the neural results that the isolation of a laughter-competent stimulus produces naturally, thanks to the activity of the brain regions and pathways that support the processing of such a stimulus and that project onto the region of the SMA. In natural laughter, the stimulus comes from within; in the case of patient A. K. it came from the tip of an electrode. In the crying patient, the electrical stimulus intervened at a later stage, well within the machinery of execution of emotion, at least one step removed from the triggering stage.

Laughter and Some More Crying

Another sort of neurological accident permits yet another glimpse at the brain stem switches of emotion. It has to do with a condition known as pathological laughter and crying. The problem has been recognized for a long time in the history of neurology, but only recently has it been possible to make sense of it in terms of brain anatomy and physiology. Patient C., whom I studied in collaboration with Josef Parvizi and Steven Anderson, provides a perfect illustration of the problem.[36]

When C. suffered a small stroke affecting the brain stem, the physician who first took care of him considered his patient lucky. Some brain stem strokes can be fatal and many leave patients with terrible disabilities. This particular stroke seemed to have caused relatively minor problems with movement, and there was a good chance those problems would abate. In that regard, C.'s condition followed the expected course. What was neither expected nor easy to deal with was a symptom that had patient, family, and caregivers at a complete loss. Patient C. would burst into the most arresting crying or the most spectacular laughter for no perceptible cause. Not only would the motive of the outburst be inapparent, but its emotional value might be diametrically opposed to the affective tenor of the moment. In the middle of a serious conversation regarding his health or finances, C. might literally burst at the seams with laughter while attempting without success to suppress it. Likewise, in the middle of the most trivial conversation, C. might sob his heart out, again unable to suppress those reactions. The outbursts might follow each other in quick succession, leaving C. barely time enough to take a breath and say that he was not in control, that neither laughter nor crying were really meant as such, that no thoughts in his mind justified such strange behavior. Needless to say, the patient was not wired to any electrical current and no one was pulling a switch on him. Yet the outcome was the same. As a result of an area of damage in the neural system constituted by nuclei in the brain stem and in the cerebellum, C. would engage these emotions without a proper mental cause and would find it difficult to rein in those emotions. No less importantly, C. would end up feeling somewhat sad or somewhat giddy, although at the start of an episode he would be neither happy nor sad, and would be having neither felicitous nor troubling thoughts. Once again, an unmotivated emotion caused a feeling and brought on a mental state consonant with the valence of a repertoire of body actions.

The fine mechanism that allows us to control laughter and crying according to the social and cognitive context has been a mystery. The study of this patient lifted part of the mystery and revealed that nuclei in the pons and in the cerebellum seem to play an important role in the control mechanism. Subsequent investigations of other patients with the same situation and comparable lesions have strengthened the conclusions. You can imagine the control mechanism as follows: Within the brain stem, systems of nuclei and pathways can be switched on to engender stereotypical laughter or crying. Then another system in the cerebellum modulates the basic laughter and crying devices. The modulation is achieved by changing, for example, the threshold for laughter or crying, the intensity and duration of some of the component movements, and so on.[37] In normal circumstances the system can be influenced by activity in the cerebral cortex—the several regions that work as an ensemble and represent on each given occasion the context in which an emotionally competent stimulus will cause a lot or very little of whatever kind of laughter or crying is appropriate. In turn, the system can influence the cerebral cortex itself.

The case of patient C. also provides a rare glimpse of the interplay between the appraisal process that precedes emotions and the actual execution of the emotions we have been considering. The appraisal can modulate the ensuing emotional state and, in turn, be modulated by it. When the appraisal and execution processes become disconnected, as they did in C., the result can be chaotic.

If the previous case reveals the dependence of behavioral and mental processes on multicomponent systems, this case reveals how those processes depend on a complicated interplay among those components. We are far away from single "centers" and far away from the idea that neural pathways work in a single direction.

From the Active Body to the Mind

The phenomena we have discussed in this chapter—emotions-proper, appetites, and the simpler regulatory reactions—occur in the theater of the body under the guidance of a congenitally wise brain designed by evolution to help manage the body. Spinoza intuited that congenital neurobiological wisdom and encapsulated the intuition in his *conatus* statements, the notion that, of necessity, all living organisms endeavor to preserve themselves without conscious knowledge of the undertaking and without having decided, as individual selves, to undertake anything. In short, they do not know the problem they are trying to solve. When the consequences of such natural wisdom are mapped back in the brain, the result is feelings, the foundational component of our minds. Eventually, as we shall see, feelings can guide a deliberate endeavor of self-preservation and assist with making choices regarding the manner in which self-preservation should take place.

Feelings open the door for some measure of willful control of the automated emotions.

Evolution appears to have assembled the brain machinery of emotion and feeling in installments. First came the machinery for producing reactions to an object or event, directed at the object or at the circumstances—the machinery of emotion. Second came the machinery for producing a brain map and then a mental image, an idea, for the reactions and for the resulting state of the organism—the machinery of feeling.

The first device, emotion, enabled organisms to respond effectively but not creatively to a number of circumstances conducive or threatening to life—"good for life" or "bad for life" circumstances, "good for life" or "bad for life" outcomes. The second device, feeling, introduced a mental alert for the good or bad circumstances and prolonged the impact of emotions by affecting attention and memory lastingly. Eventually, in a fruitful combination with past memories, imagination, and reasoning, feelings led to the emergence of foresight and the possibility of creating novel, nonstereotypical responses.

As is often the case when new devices are added, nature used the machinery of emotion as a start and tinkered a few more components. In the beginning was emotion, but at the beginning of emotion was action.

CHAPTER 3

Feelings

What Feelings Are

In my attempt to explain what feelings are, I will begin by asking the reader a question: When you consider any feeling you have experienced, pleasant or not, intense or not, what do you regard as the contents of that feeling? Note that I am not inquiring about the cause of the feeling; or about the intensity of the feeling; or about its positive or negative valence; or about what thoughts came into your mind in the wake of the feeling. I really mean the mental contents, the ingredients, the stuff that makes a feeling.

In order to get this thought experiment going let me offer some suggestions: Think of lying down in the sand, the late-day sun gently warming your skin, the ocean lapping at your feet, a rustle of pine needles somewhere behind you, a light summer breeze blowing, 78 degrees F and not a cloud in the sky. Take your time and savor the experience. I will assume you were not bored to tears and that instead you felt very well, exceedingly well, as a friend of mine likes to put it, and the question is, what did that "feeling well" consist of? Here are just a few clues: Perhaps the warmth of your skin was comfortable. Your breathing was easy, in and out, unimpeded by any resistance in the chest or at the throat. Your muscles were so relaxed that you could not sense any pull at the joints. The body felt light, grounded but airy. You could survey the organism as a whole and you could sense its machinery working smoothly, with no glitches, no pain, simple perfection. You had the energy to move, but somehow you preferred to remain quiet, a paradoxical combination of the ability and inclination to act and the savoring of the stillness. The body, in brief, felt different along a number of dimensions. Some dimensions were quite apparent, and you actually could identify their locus. Others were more elusive. For example, you felt well-being and an absence of pain, and although the locus of the phenomenon was the body and its operations, the sensation was so diffuse that it was difficult to describe precisely where that was happening in the body.

And there were mental consequences of the state of being just described. When you could direct your attention away from the sheer well-being of the moment, when you could enhance the mental representations that did not pertain directly to your body, you found that your mind was filled with thoughts whose themes created a new wave of pleasurable feeling. The picture of events you eagerly anticipated as pleasurable came into mind, as did scenes you enjoyed experiencing in the past. Also, you found that your cast of mind was, well, felicitous. You had adopted a mode of thinking in which images had a sharp focus and flowed abundantly and effortlessly. There were two consequences for all that good feeling. The appearance of *thoughts with themes consonant with the emotion*; and *a mode of thinking, a style of mental processing*, which increased the speed of image generation and made images more abundant. You had, as Wordsworth did up in Tintern Abbey, "sensations sweet felt in the blood and felt along the heart" and found that these sensations were "passing even into [your] purer mind in tranquil restoration." What you usually regard as "body" and as "mind" blended in harmony. Any conflicts now seemed abated. Any opposites now seemed less opposite.

I would say that what defined the pleasurable feeling of those moments, what made the feeling deserve the distinctive term feeling and be different from any other thought, was the mental representation of parts of the body or of the whole body as operating in a certain manner. Feeling, in the pure and narrow sense of the word, was *the idea of the body being in a certain way*. In this definition you can substitute idea for "thought" or "perception." Once you looked beyond the object that caused the feeling and the thoughts and mode of thinking consequent to it, the core of the feeling came into focus. Its contents consisted of representing a particular state of the body.

The same comments would apply entirely to feelings of sadness, feelings of any other emotion, feelings of appetites, and feelings of any set of regulatory reactions unfolding in the organism. Feelings, in the sense used in this book, arise from any set of homeostatic reactions, not just from emotions-proper. They translate the ongoing life state in the language of the mind. I propose that there are distinctive "body ways" resulting from different homeostatic reactions, from simple to complex. There also are distinctive causative objects, distinctive consequent thoughts, and consonant modes of thinking. Sadness, for example, is accompanied by low rates of image production and hyperattentiveness to images, rather than by the rapid image change and short attention span that goes with high happiness. Feelings are perceptions, and I propose that the most necessary support for their perception occurs in the *brain's body maps*. These maps refer to parts of the body and states of the body. Some variation of pleasure or pain is a consistent content of the perception we call feeling.

Alongside the perception of the body there is the perception of thoughts with themes consonant with the emotion, and a perception of a certain mode of thinking, a style of mental processing.

How does this perception come about? It results from constructing metarepresentations of our own mental process, a high-level
operation in which a part of the mind represents another part of
the mind. This allows us to register the fact that our thoughts
slow down or speed up as more or less attention is devoted to
them; or the fact that thoughts depict objects and events at close
range or at a distance. My hypothesis, then, presented in the
form of a provisional definition, is that *a feeling is the perception of
a certain state of the body along with the perception of a certain mode
of thinking and of thoughts with certain themes*. Feelings emerge
when the sheer accumulation of mapped details reaches a certain
stage. Coming from a different perspective, the philosopher
Suzanne Langer captured the nature of that moment of emergence by saying that when the activity of some part of the nervous system reaches a "critical pitch" the process is felt.[1] Feeling is
a consequence of the ongoing homeostatic process, the next step
in the chain.

The above hypothesis is not compatible with the view that the
essence of feelings (or the essence of emotions when emotions
and feelings are taken as synonyms) is a collection of thoughts
with certain themes consonant with a certain feeling label, such
as thoughts of situations of loss in the case of sadness. I believe
the latter view empties the concept of feeling hopelessly. If feelings were merely clusters of thoughts with certain themes, how
could they be distinguished from any other thoughts? How would
they retain the functional individuality that justifies their status as
a special mind process? My view is that feelings are functionally
distinctive because their essence consists of the thoughts that represent the body involved in a reactive process. Remove that
essence and the notion of feeling vanishes. Remove that essence
and one should never again be allowed to say "I feel" happy, but

rather, "I think" happy. But that begs a legitimate question: What makes thoughts "happy"? If we do not experience a certain body state with a certain quality we call pleasure and that we find "good" and "positive" within the framework of life, we have no reason whatsoever to regard any thought as happy. Or sad.

As I see it, the *origin* of the perceptions that constitute the essence of feeling is clear: There is a general object, the body, and there are many parts to that object that are continuously mapped in a number of brain structures. The *contents* of those perceptions also are clear: varied body states portrayed by the body-representing maps along a range of possibilities. For example, the micro- and macrostructure of tensed muscles are a different content than relaxed muscles. The same is true of the state of the heart when it beats fast or slow, and for the function of other systems—respiratory, digestive—whose business can proceed quietly and harmoniously, or with difficulty and poor coordination. Another example, and perhaps the most important one, is the composition of the blood relative to some chemical molecules on which our life depends, and whose concentration is represented, moment by moment, in specific brain regions. The particular state of those body components, as portrayed in the brain's body maps, is a content of the perceptions that constitute feelings. The immediate *substrates* of feelings are the mappings of myriad aspects of body states in the sensory regions designed to receive signals from the body.

Someone might object that we do not seem to register consciously the perception of all those body-part states. Thank goodness we do not register them all, indeed. We do experience some of them quite specifically and not always pleasantly—a disturbed heart rhythm, a painful contraction of the gut, and so forth. But for most other components, I hypothesize that we experience them in "composite" form. Certain patterns of internal milieu chemistry, for example, register as background feelings of energy, fatigue, or malaise. We also experience the set of behavioral changes that become appetites and cravings. Obviously we do not "experience" the blood level of glucose dropping below its lower admissible threshold, but we rapidly experience the consequences of that drop: certain behaviors are engaged (e.g., appetite for food); the muscles do not obey our commands; we feel tired.

Experiencing a certain feeling, such as pleasure, is perceiving the body as being in a certain way, and perceiving the body in whatever way requires sensory maps in which neural patterns are instantiated and out of which mental images can be derived. I caution that the emergence of mental images from neural patterns is not a fully understood process (there is a gap in our understanding that we review in chapter 5). But we know enough to hypothesize that the process is supported by identifiable substrates—in the case of feelings, several maps of body state in varied brain regions—and subsequently involves complex interactions among regions. The process is not localized to one brain area.

In brief, the essential content of feelings is the mapping of a particular body state; the substrate of feelings is the set of neural patterns that map the body state and from which a mental image of the body state can emerge. A feeling in essence is an idea—an idea of the body and, even more particularly, an idea of a certain aspect of the body, its interior, in certain circumstances. A feeling of emotion is an idea of the body when it is perturbed by the emoting process. As we shall see in the pages ahead, however, the mapping of the body that constitutes the critical part of this hypothesis is unlikely to be as direct as William James once imagined.

Is There More to Feelings than the Perception of Body State?

When I say that feelings are largely constituted by the perception of a certain body state or that the perception of a body state forms the essence of a feeling, my use of the words largely and essence is not casual. The reason for the subtlety can be gleaned from the hypothesis-definition of feeling we have been discussing. In many circumstances, especially when there is little or no time to examine feelings, feelings are solely the perception of a certain body state. In other circumstances, however, feeling involves the perception of a certain body state *and* the perception of a certain accompanying *mind state*—the changes in mode of thinking to which I referred earlier as part of the consequences of feeling. What happens in those circumstances is that as we hold images of the body as being such and so, and, in parallel, we hold images of our own thinking style.

In certain circumstances of feeling, in the most advanced variety of the phenomenon perhaps, the process is anything but simple. It encompasses the following: the body states that are the essence of the feeling and give it a distinctive content; the altered mode of thinking that accompanies the perception of that essential body state; and the sort of thoughts that agree, in terms of theme, with the kind of emotion being felt. On those occasions, if you take the example of a positive feeling, we might say that the mind represents more than well-being. The mind also represents well-thinking. The flesh operates harmoniously, or so the mind says, and our thinking powers are either at the top of their game or can be taken there. Likewise, feeling sad is not just about a sickness in the body or about a lack of energy to continue. It is often about an inefficient mode of thought stalling around a limited number of ideas of loss.

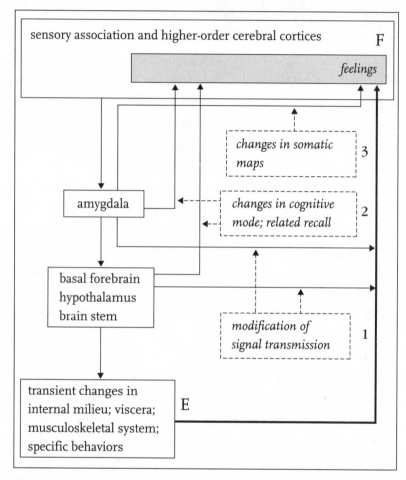

Figure 3.1: The continuation of the diagram in figure 2.5, now leading all the way to feelings of fear. The transmission of signals from the body to the brain (the outer arrow moving from box E at bottom left to box F in upper right) can be influenced by the triggering and execution sites (arrows from box 1 marked modification of signal transmission*). The triggering and execution sites also influence the process by creating* changes in cognitive mode and related recall *(box 2), and by making direct* changes in somatic maps *(box 3) that constitute the proximate neural substrate for feelings. Note that both the appraisal/evaluation stage and the final feelings stage occur at the cerebral level, within sensory association and higher-order cerebral cortices.*

Feelings Are Interactive Perceptions

Feelings are perceptions and, in some ways, they are comparable to other perceptions. For example, actual visual perceptions correspond to external objects whose physical characteristics impinge on our retinas and temporarily modify the patterns of sensory maps in the visual system. Feelings also have an object at the origin of the process and the physical characteristics of the object also prompt a chain of signals that transit through maps of the object inside the brain. Just as in the case of visual perception, there is a part of the phenomenon that is due to the object, and a part that is due to the internal construction the brain makes of it. But something that is different—and the difference is not trivial—is that, in the case of feelings, the *objects and events at the origin* are well inside the body rather than outside of it. Feelings may be just as mental as any other perception, but the objects being mapped are parts and states of the living organism in which feelings arise.

This important difference begets two others. First, in addition to being linked to an object at the origin—the body—feelings also are linked to the emotionally competent object that initiated the emotion-feeling cycle. In a curious way the emotionally competent object is responsible for establishing the object at the origin of a feeling. Thus when we refer to the "object" of an emotion or of a feeling we must qualify the reference and make clear which object we mean. *The sight of a spectacular seascape is an emotionally competent object. The body state that results from beholding that seascape x is the actual object at the origin x, which is then perceived in the feeling state.*

Second, and no less importantly, the brain has a direct means to respond to the object as feelings unfold because the object at the origin is inside the body, rather than external to it. The brain can act directly on the very object it is perceiving. It can do so by modifying the state of the object, or by altering the transmission of signals from it. The object at the origin on the one hand, and the brain map of that object on the other, can influence each other in a sort of reverberative process that is not to be found, for example, in the perception of an external object. You can look at Picasso's *Guernica* as intensely as you wish, for as long as you wish, and as emotionally as you wish, but nothing will happen to the painting itself. Your thoughts about it change, of course, but the object remains intact, one hopes. In the case of feeling, the object itself can be changed radically. In some instances the changes may be akin to taking a brush and fresh paint and modifying the painting.

In other words, feelings are not a passive perception or a flash in time, especially not in the case of feelings of joy and sorrow. For a while after an occasion of such feelings begins—for seconds or for minutes—there is a dynamic engagement of the body, almost certainly in repeated fashion, and a subsequent dynamic variation of the perception. We perceive a series of transitions. We sense an interplay, a give and take.[2]

At this point, you might object to my wording and say that the arrangement I am describing applies to feelings of emotion and related regulatory phenomena but perhaps not to other kinds of feeling. And I would have to say that the only other proper usage of the term feeling pertains to the act of touching or to its result, a tactile perception. In reference to the dominant use of the term feeling, as agreed at the outset, I would say that *all* feelings are feelings of some of the basic regulatory reactions we discussed earlier, or of appetites, or of emotions-proper, from straight pain to beatitude. When we talk of the "feeling" of a certain shade of blue or of the "feeling" of a certain musical note, we actually are referring to the affective feeling that accompanies our seeing that shade of blue or hearing the sound of that note, regardless of how subtle the aesthetic perturbation may be.[3] Even when we some-what misuse the notion of feeling—as in "I feel I am right about this" or "I feel I cannot agree with you"—we are referring, at least vaguely, to the feeling that accompanies the idea of believing a certain fact or endorsing a certain view. This is because believing and endorsing *cause* a certain emotion to happen. As far as I can fathom, few if any perceptions of any object or event, actually present or recalled from memory, are ever neutral in emotional terms. Through either innate design or by learning, we react to most, perhaps all, objects with emotions, however weak, and sub-sequent feelings, however feeble.

Mixing Memory with Desire: An Aside

Over the years, I often have heard it said that perhaps we can use the body to explain joy, sorrow, and fear, of course, but certainly not desire, love, or pride. I am always intrigued by this reluctance and whenever the statement is made directly to me I always reply in the same vein: Why not? Let me try. It makes no difference whether my debater is a man or a woman, I always propose the same thought experiment: Consider the time, I hope recently, when you saw a woman or man (fill in your preference) who awoke in you, within a matter of seconds, a distinct state of lust. Try to think through what took place, in physiological terms, using the neurobiological devices I have been discussing.

The object of origin for that awakening presented itself, in all its glory, probably not whole but in parts. Maybe what first arrested your attention was the shape of an ankle, how it connected with the back of a shoe and how it dissolved into a leg, no longer seen but just imagined, under a skirt. ("She came at me in sections; she had more curves than a scenic highway," said Fred Astaire, describing the arrival of the tantalizing Cyd Charisse in *The Bandwagon*.) Or maybe it was the shape of a neck sticking up from a shirt. Or maybe it was not a part at all but the carriage, moves, energy, and resolve that propelled a whole body forward. Whatever the presentation, the appetite system was engaged and the appropriate responses were selected. What made up those responses? Well, preparations and simulations, as it turns out. The appetite system promoted a number of subtle and perhaps not-so-subtle body changes that are part of a routine of readiness for the eventual consummation of the appetite. Never mind that, in civilized company, consummation may never come. There were rapid chemical alterations in your internal milieu, changes in heartbeat and respiration compatible with your barely defined wishes, redistributions of blood flow, and a muscular presetting of the varied patterns of motion that you might possibly engage in, but probably would not. The tensions in your musculoskeletal system were rearranged, in fact, tensions arose where there were none, just moments ago, and odd relaxations showed up as well. Adding to all this, imagination kicked in, making wishes more clear now. The machinery of reward, chemical and neural, was in full swing, and the body deployed some of the behaviors associated with the eventual feeling of pleasure. Very stirring, indeed, and very mappable in the body-sensing and cognitive support brain regions. Thinking of the goal of the appetite caused pleasant emotions and the corresponding pleasant feelings. Desire was now yours.

In this example, the subtle articulation of appetites, emotions, and feelings becomes apparent. If the goal of the appetite was permissible and fulfilled, the satisfaction would cause a specific emotion of joy, maybe, one hopes, and turn the feeling of desire into the feeling of elation. If the goal was thwarted, anger might ensue, instead. But if the process stayed in suspension for a while, in the delicious never-never land of daydreams, it eventually would die down, quietly. Sorry, no cigarette afterward. You are not inside a film noir.

Are hunger and thirst that different from sexual desire? Simpler, no doubt, but not really different in mechanism. That is the reason why all three can blend so easily and at times even compensate mutually. The main distinction comes from memory, I would say, from the manner in which the recall and permanent rearrangement of our personal experiences play a role in the unfolding of desire, more so than they usually do in hunger or thirst. (But let us beware of gastronomes and wine connoisseurs who will disabuse us of this idea.) Be that as it may, there is a rich interplay between the object of desire and a wealth of personal memories pertinent to the object—past occasions of desire, past aspirations, and past pleasures, real or imagined.

Are attachment and romantic love amenable to comparable biological accounts? I do not see why not, provided the attempt to explain fundamental mechanisms is not pushed to the point of explaining unnecessarily one's unique personal experiences and trivializing the individual. We can certainly separate sex from attachment, thanks to the investigation of how two hormones we regularly manufacture in our bodies, the peptides oxytocin and vasopressin, affect the sexual and attachment behavior of a charming species, the prairie voles. Blocking oxytocin in a female prairie vole prior to mating does not interfere with sexual behavior but preempts her attachment to the sexual partner. Sex yes, fidelity no. Blocking vasopressin in the male prairie vole prior to mating has a comparable effect. Mating still takes place, but the usually faithful male prairie vole does not bond to the female, nor does he bother to protect his date and, eventually, their progeny.[4] Sex and attachment are not romantic love, of course, but they are part of its genealogy.[5]

The same goes for pride or shame, two affects that often have been said not to relate to body expressions at all. But of course they do. Can one imagine a more distinct body posture than that of the person beaming with pride? What exactly *beams*? The eyes to be sure, wide open, focused and intent on taking on the world; the chin held high; the neck and torso as vertical as they can get; the chest unfearingly filled with air; the steps firm and well planted. These are just some body changes we can *see*. Compare them with those of the shamed and humiliated man. To be sure, the emotionally competent situation is quite different for shame. The thoughts that accompany this emotion and are subsequent to the onset of the feelings are as different as day and night. But here too we find an entirely distinct and mappable state between the triggering event and the consonant thoughts.

And so it must be for brotherly love, the most redeeming of all feelings, a feeling that depends for its modulation on the unique repository of autobiographical records that define our identities. Yet it still rests, as Spinoza so clearly gleaned, on occasions of pleasure—bodily pleasure, what else?—prompted by thoughts of a particular object.

Feelings in the Brain: New Evidence

The notion that feelings are related to neural mappings of body state is now being put to experimental test. Recently we conducted an investigation of the patterns of brain activity that occur in association with feeling certain emotions.[6] The hypothesis that guided the work stated that when feelings occur there is a significant engagement of the areas of the brain that receive signals from varied parts of the body and thus map the ongoing state of the organism. Those brain areas, which are placed at several levels of the central nervous system, include the cingulate cortex; two of the somatosensory cortices (known as the insula and SII); the hypothalamus; and several nuclei in the brain stem tegmentum (the back part of the brain stem).

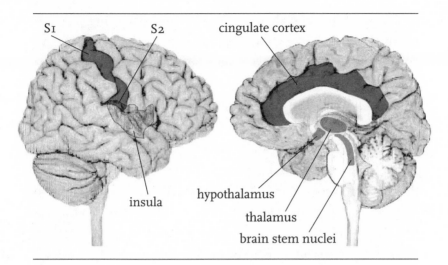

Figure 3.2: The main somatosensing regions, from the level of the brain stem to the cerebral cortex. Normal feelings of emotion require the integrity of all these regions, but the role that each region plays in the process is different. All regions are important but some regions (insula, cingulate cortex, and brain stem nuclei) are more important than others. The quietly hidden insula may be the most important of all.

To test the hypothesis, my colleagues Antoine Bechara, Hanna Damasio, and Daniel Tranel and I enlisted the cooperation of more than forty people evenly divided by gender. None had ever suffered from neurological or psychiatric disease. We told the group we wished to study the patterns of activity in their brains while they experienced one of four possible feelings: happiness, sadness, fear, or anger.

The investigation depended on measuring the amount of blood flow in multiple brain areas using a technique known as PET (for positron-emission tomography). It is known that the amount of blood flowing into any region of the brain is closely correlated with the metabolism of the neurons in that region, and the metabolism, in turn, correlates with the amount of local activity of the neurons. In the tradition of this technique, statistically significant increases or decreases in blood flow within a certain region indicate that neurons in the region were disproportionately active or inactive during the performance of a given mental task.

The key to this experiment was finding a way to trigger the emotions. We asked each subject to think of an emotional episode from their lives. The only requirement: The episode had to be especially powerful and involve happiness, sadness, fear, or anger. We then asked each subject to think in great detail about the specific episode, and bring forth all the imagery they could so that the emotions of that past event could be reenacted as intensely as possible. As previously noted, this sort of emotional memory device is a mainstay of some acting techniques, and we were pleased to find that the device worked in the context of our experiment as well. Not only have most adults experienced these episodes, but, as it turns out, most also can conjure up fine details and literally relive those emotions and feelings with surprising intensity.

We asked each subject to think of an emotional episode from their lives. All that was required was that the episode would have been especially powerful.

During a preexperiment phase we determined which emotions each subject could reenact best, and we measured such physiological parameters as heart rate and skin conductance during the reenactment. Then we began the actual experiment. We asked each subject to reenact an emotion—say, sadness—and he or she started the process of imagining the particular episode in the quiet of the scanning room. The subjects had been instructed to signal with a small hand movement at the moment they began feeling the emotion, and it was only after this signal that we started collecting data on brain activity. The experiment was skewed toward measuring the brain activity during actual feeling, rather

than during the earlier stage of recalling an emotionally competent object and triggering an emotion.

The analysis of the data offered ample support for our hypothesis. All the body-sensing areas under scrutiny—the cingulate cortex, the somatosensory cortices of insula and SII, the nuclei
in the brain stem tegmentum—showed a statistically significant

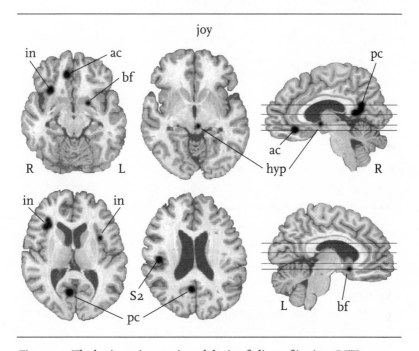

*Figure 3.3: The brain regions activated during feelings of joy in a PET
experiment. The two panels on the right of the figure show a medial (internal)
view of the right hemisphere (top) and the left hemisphere (bottom). There
are significant changes in activity in the anterior cingulate (ac), posterior
cingulate (pc), the hypothalamus (hyp), and the basal forebrain (bf). The
four panels on the left depict the brain in axial (near horizontal) slices. The
right hemisphere is marked R and the left L. Note the significant activity in
the region of the insula (in), shown in two slices and in both right and left
hemispheres, and in the posterior cingulate (pc), also shown in two slices.*

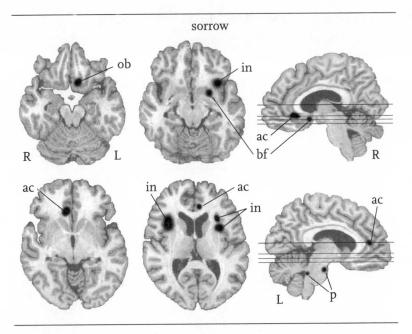

Figure 3.4: Brain maps from the same experiment corresponding to feelings of sadness. There is significant activity in the insula (in), again in both hemispheres and in more than one slice, and it differs from the condition of joy. The same applies to the significant changes in the anterior cingulate.

pattern of activation or deactivation. This indicated that the mapping of body states had been significantly modified during the process of feeling. Moreover, as we expected, these patterns of activation or deactivation varied among the emotions. In the same way that one can sense that our bodies are differently conformed during the feeling of joy or sadness, we were able to show that the brain maps corresponding to those body states were different as well.

These findings were important on many counts. It was gratifying to find that feeling an emotion was indeed associated with changes in the neural mapping of body state. More importantly, we now had a more solid pointer as to where to look in future studies of the neurobiology of feeling. The results told us in no uncertain terms that some of the mysteries of the physiology of feelings could be solved in the neural circuitry of body-sensing brain regions and in the physiological and chemical operation of those circuitries.

The study also provided some unexpected and welcome results. We had monitored the subjects' physiological responses continuously and were able to note that *the changes in skin conductance always preceded the signal that a feeling was being felt.* In other words, the electrical monitors registered the seismic activity of emotion unequivocally *before* the subjects moved their hand to indicate the experience had begun. Although we had not planned to look into this issue, the experiment offered still more evidence that emotional states come first and feelings after.

Another suggestive result had to do with the state of the regions of cerebral cortex that are related to the thought process, namely, the cortices of the lateral and polar aspects of the frontal lobe. We had not formulated a hypothesis to explain how the modes of thinking that are engaged differently in the various feelings would reveal themselves in the brain. Yet the findings were quite sensible. In the sadness condition there were marked deactivations in prefrontal cortices (to a considerable extent this suggests a reduction of activity in the overall region). In the happiness condition, we found the opposite (a considerable indication of increased activity in the region). These findings accord well with the fact that the fluency of ideation is reduced in sadness and increased in happiness.

A Comment on Related Evidence

It is always pleasant to find evidence in favor of one's theoretical preferences, but one should not be too encouraged by one's own findings until corroborating evidence is found. If the strong pointer to somatosensing regions that we encountered in our feelings study is a solid fact, others should find compatible evidence. Indeed, abundant compatible evidence is now on record, based on the same approach (functional imaging techniques such as PET and fMRI), and pertaining to an eclectic collection of feelings.

The studies by Raymond Dolan and his colleagues are especially pertinent here because they addressed our work specifically, although unrelated work also has produced compatible results.[7] Whether the participant is experiencing the pleasures of eating chocolate or the insane feeling of romantic love, the guilt of Clytemnestra, or the excitement of erotic film excerpts, the key areas targeted in our experiments (e.g., the insular cortex and the cingulate cortex) exhibit significant changes. Those areas are either more active or less, in varied patterns within the key region, testifying to the notion that states of feeling are correlated with significant engagement of these brain regions.[8] Predictably other regions are involved as well, namely regions involved in the actual generation of the related emotions, but the point to be made here is that changed activity in the somatosensing regions is correlated with feeling states. As we will see later in this chapter, the feelings associated with taking narcotics or craving them also result in significant engagement of the same somatosensing areas.

There is an intimate and telling three-way connection between certain kinds of music, feelings of either great sorrow or great joy, and the body sensations we describe as "chills" or "shivers" or "thrills." For curious reasons, certain musical instruments, particularly the human voice, and certain musical compositions, evoke emotive states that include a host of skin responses such as making the hair stand on end, producing shudders, and blanching the skin.[9] Perhaps nothing is more illustrative for our purposes than evidence from a study conducted by Anne Blood and Robert Zatorre. They wanted to study neural correlates of pleasurable states caused by listening to music capable of evoking chills and shivers down the spine.[10] The investigators found those correlates in the somatosensing regions of the insula and anterior cingulate, which were significantly engaged by musically thrilling pieces. Moreover, the investigators correlated the intensity of the activation with the reported thrill value of the pieces. They demonstrated that the activations were related to the thrilling pieces (which individual participants handpicked) and not to the mere presence of music. Curiously, on other grounds, it is suspected that the appearance of chills is caused by the immediate availability of endogenous opioids in the brain regions modified by these feelings.[11] The study also identified regions involved in producing the emotive responses behind the pleasurable states—e.g., right orbitofrontal cortices, left ventral striatum, and regions that were negatively correlated with the pleasurable state—e.g., right amygdala—much as our own study did.

Studies of pain processing also speak to the issue. In a telling experiment conducted by Kenneth Casey, participants were subjected to hand pain (their hands were immersed in ice-cold water) or a nonpainful vibratory stimulus in the hand while their brains were scanned.[12] The pain condition resulted in notable changes of activity in two somatosensing regions (insula and SII). The vibratory condition produced activation of another somatosensing region (SI), but not of the insula and SII, the regions most closely aligned with feelings of emotion. After each condition researchers gave the patients fentanyl (a drug that mimics morphine because it acts on μ [mu] opioid receptors) and scanned the subjects once again. In the pain condition, fentanyl managed to reduce *both* the pain *and* the engagement of the *insula* and *SII*. In the vibratory condition the administration of fentanyl left intact

both the perception of vibration *and* the activation of *SI*. These results reveal with some clarity a separate physiological arrangement for feelings that relates to pain or pleasure and for "feelings"
of tactile or vibratory sensations. Insula and SII are strongly associated with the former, SI with the latter. Elsewhere I have noted
that the physiological support of emotion and pain sensation can
be dissociated by drugs such as Valium, which remove the affect
component of pain but leave the sensation of pain intact. The apt
description for such situations is that you "feel" pain but do
not care.[13]

Some More Corroborating Evidence

It has been shown convincingly that the feeling of thirst is associated with significant changes of activity in the cingulate cortex and in the insular cortex.[14] The state of thirst itself results
from detecting a water imbalance and from the subtle interplay
between hormones such as vasopressin and angiotensin II and
regions of the brain such as the hypothalamus and periaqueductal gray, whose job it is to call into action thirst-relief behaviors, a
collection of highly coordinated hormonal releases and motor
programs.[15]

I will spare the reader the description of how feeling the urge
to empty one's bladder, male or female, or the feeling of having
emptied it, are correlated with changes in, yes, the cingulate cortex.[16] But I should say something about appetites and desires
aroused by viewing erotic films. Predictably, the cingulate cortex
and the insular cortex are very much engaged so that we can feel
the excitement. Regions such as the orbitofrontal cortices and the
striatum are engaged as well—they are whipping up the excitement, actually. When it comes to the gender of the participants,
however, there is a remarkable difference in the engagement of

one region, the hypothalamus. Males engage the area significantly; females do not.[17]

The Substrate of Feelings

When David Hubel and Torsten Wiesel began their celebrated
work on the neural basis of vision in the 1950s there was no
inkling yet of the kind of organization they would discover within
the primary visual cortex, namely the kind of submodular organization that permits us to construct maps related to a visual object.[18] The means behind the visual mappings was a mystery. On
the other hand, there was a perfect inkling as to the general area
where the secrets should be searched for, namely, the chain of
pathways and processing stations that began in the retina and
abutted in the visual cortices. When we consider the field of feeling today it is apparent that we have just reached a stage comparable, in many regards, to that of vision research at the time
Hubel and Wiesel launched their program. Until recently, many
scientists have been reluctant to accept that the somatosensory
system could be a critical substrate of feeling. This is perhaps the
last remnant of resistance to William James's conjecture that
when we feel emotions we perceive body states. It also is a bizarre
accommodation to the idea that affective feelings might not have
a sensory base comparable to that of vision or hearing. The evidence from lesion studies and more recently from the functional
imaging studies cited earlier has now changed this acquiescence
irrevocably. Yes, somatosensing regions are involved in the feeling process, and yes, a major partner in the firm of somatosensory cortices, the insula, is involved perhaps more significantly
than any other structure. SII, SI, and the cingulate cortex also are
involved, but their participation is at a different level. For a variety
of reasons I believe the involvement of the insula is pivotal.

The above facts bring together two strands of evidence: From the introspective analysis of feeling states, it stands to reason that feelings ought to depend on somatosensory processing. From neurophysiological/imaging evidence, a structure such as the insula is indeed differentially engaged in feeling states, as we have just seen.[19]

But another strand of recent evidence makes this convergence even more powerful. It so happens that the peripheral nerve fibers and neural pathways dedicated to conveying information from the body's interior to the brain do *not* terminate, as once thought, in the cortex that receives signals related to touch (SI, the primary somatosensory cortex). Instead, those pathways terminate in their own dedicated region, the *insular cortex* itself, precisely the same region whose activity patterns are perturbed by feelings of emotion.[20]

The neurophysiologist and neuroanatomist A. D. Craig has uncovered important evidence and deserves great credit for pursuing an idea lost in the mists of early neurophysiology and traditionally denied in textbook neurology—the idea that we are privy to a sense of the body's interior, an *interoceptive* sense.[21] In other words, the very same region that both theoretical proposals and functional imaging studies relate to feelings turns out to be the recipient of the class of signals most likely to represent the content of feelings: signals related to pain states; body temperature; flush; itch; tickle; shudder; visceral and genital sensations; the state of the smooth musculature in blood vessels and other viscera; local pH; glucose; osmolality; presence of inflammatory agents; and so forth. From a variety of perspectives, then, the somatosensing regions appear to be a critical substrate for feelings, and the insular cortex appears to be the pivotal region of the set. This notion, no longer a mere hypothesis, constitutes a platform from which a new level of inquiry can be launched into the finer neurobiology of feelings in the years ahead.

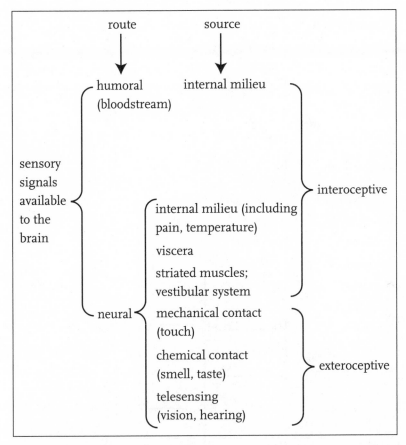

Figure 3.5A: A key to the kinds of sensory signal received by the brain. There are two routes of transmission: humoral *(in which, for example, chemical molecules conveyed by the bloodstream directly activate neural sensors in the hypothalamus or in circumventricular organs such as the area postrema); and* neural *(in which electrochemical signals are transmitted in neural pathways by the axons of neurons firing upon the cell body of other neurons, across synapses). There are two sources for all these signals: the external world (exteroceptive signals), and the inner world of the body (interoceptive signals). Emotions are, by and large, modifications of the inner world. Thus the sensory signals that constitute the basis for feelings of emotion are largely interoceptive. The main source of those signals is the viscera and the internal milieu, but signals related to the state of the musculoskeletal and vestibular systems participate as well.*[22]

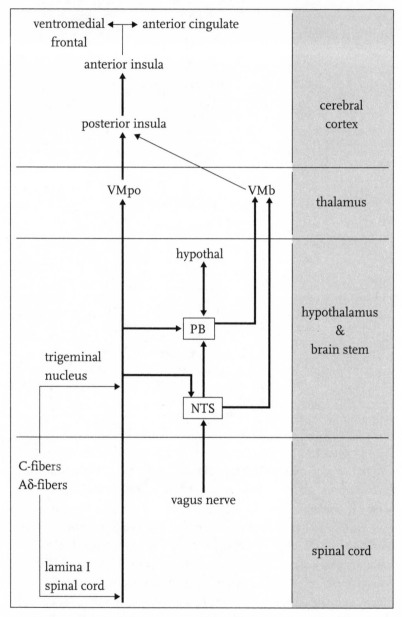

Figure 3.5B: Signaling from body to brain. A diagram of the critical structures involved in conveying internal milieu and visceral signals to the brain. A substantial part of the critical signaling is conveyed by pathways from the spinal cord and the brain stem's trigeminal nucleus. At every level of the spinal cord, in a region known as "lamina I" (in the posterior horn

Who Can Have Feelings?

In attempting to discover the basic processes that permit feeling, one comes to the following considerations. First, an entity capable of feeling must be an organism that not only has a body but also a means to represent that body inside itself. We can think of elaborate organisms such as plants that clearly are alive and have a body, but have no means of representing parts of their bodies and the states of those parts in the sort of maps our brains provide. Plants react to many stimuli—to light, heat, water, and nutrients. Some people with green fingers even believe they react to kind words of encouragement. But they seem to lack the possibility of

---

of the spinal cord's gray matter, and in the caudal part of the trigeminal nucleus), the information conveyed by peripheral nerve fibers of the C and Aδ types (thin, unmyelinated, and slow conducting) is brought to the central nervous system. This information hails from literally everywhere in our entire body and relates to parameters as diverse as the state of contraction of smooth muscles in arteries, the amount of local blood flow, local temperature, the presence of chemicals signifying injury to local tissue, the level of pH, O_2, and CO_2. All of this information is further conveyed to a dedicated nucleus of the thalamus (VMpo) and then on to neural maps in the posterior and anterior insula. Subsequently the insula can signal to regions such as the ventromedial prefrontal cortex and the anterior cingulate cortex. On the way to the thalamus, this information is also made available to the nucleus tractus solitarius (NTS), which receives signals from the vagus nerve (a major path for information from the viscera that bypasses the spinal cord); to the parabrachial nucleus (PB); and to the hypothalamus (hypothal). The PB and the NTS, in turn, also convey signals to the insula via yet another thalamic nucleus (VMb). Intriguingly, the pathways related to the movement of the body and to its position in space use an entirely different chain of transmission. The peripheral nerve fibers that convey those signals (Aβ) are thick and conduct at fast speeds. The parts of the spinal cord and trigeminal nerve nucleus used for body movement signaling are also different, and so are the thalamic relay nuclei and the ultimate cortical target (the somatosensory cortex I).

being conscious of a feeling. The first requirement for feeling, then, comes down to the presence of a nervous system.

Second, that nervous system must be able to map body structures and body states and transform the neural patterns in those maps into mental patterns or images. Without the latter step, the nervous system would map the body changes that are the substrate of feelings without quite getting to the point of producing the idea that we call feeling.

Third, the occurrence of a feeling in the traditional sense of the term requires that its contents be known to the organism, i.e., consciousness is a requirement. The relation between feeling and consciousness is tricky. In plain terms, we are not able to feel if we are not conscious. But it so happens that the machinery of feeling is itself a contributor to the processes of consciousness, namely to the creation of the self, without which nothing can be known. The way out of the difficulty comes from realizing that the process of feeling is multitiered and branched. Some of the steps necessary to produce a feeling are the very same necessary to produce the protoself, on which self and eventually consciousness depend. But some of the steps are specific to the set of homeostatic changes being felt, i.e., specific to a certain object.

Fourth, the brain maps that constitute the basic substrate of feelings exhibit patterns of body state that have been executed under the command of other parts of the very same brain. In other words, the brain of an organism that feels creates the very body states that evoke feelings as it reacts to objects and events with emotions or appetites. In organisms capable of feeling, then, the brain is a double necessity. To be sure, it must be there to provide body mappings. Even before that, however, the brain must have been there either to command or construct the particular emotional body state that ends up being mapped as feeling.

These circumstances call attention to one likely reason why feelings became possible in evolution. Feelings probably became possible because there were brain maps available to represent body states. Those maps became possible because the brain machinery of body regulation required them in order to make its regulatory adjustments, namely those adjustments that occur during the unfolding of an emotional reaction. That means feelings depend not just on the presence of a body and brain capable of body representations, they also depend on the prior existence of the brain machinery of life regulation, including the part of the life-regulating mechanism that causes reactions such as emotions and appetites. Without the prior existence of the brain machinery behind emotions there might be nothing interesting to feel. Once again, in the beginning was emotion and its underpinnings. Feeling is not a passive process.

Body States versus Body Maps

The outline of the proposal I have presented so far is simple enough. But this is the time to make the problem more complicated. By way of background let me introduce two issues.

Our hypothesis is that whatever we feel must be based on the activity pattern of the body-sensing brain regions. If those body-sensing regions were not available to us we would not feel anything, in the same way we would not see anything if we were deprived of the key visual regions of our brain. The feelings we experience come courtesy of body-sensing regions. This may sound a bit too obvious, however I must recall that until quite recently science studiously avoided the assignment of feelings to *any* brain system; feelings were just out there, vaporously hanging in or around the brain. But now comes a potential reservation that deserves all one's attention because it is sensible but not valid. In many instances, the body-sensing regions produce a precise map of what is occurring in the body, yet in some instances they do not for the simple reason that either the activity in the mapping regions or the signals coming toward them may have been modified

in some way. The mapped pattern has lost fidelity. Does this compromise the idea that we feel what is mapped in the body-sensing
brain? It does not. More about this in an instant.

The second issue regards William James, who proposed that
feelings are necessarily a perception of the actual body changed by
emotion. One of the reasons why James's insightful conjecture was
attacked and eventually abandoned for a long time had to do with
the notion that, somehow, making feelings dependent on the perception of actual body states would delay the process of feeling, rendering it ineffectual. It does take time to change the body and map
the consequent changes. As it happens, however, it also takes quite
a time to feel. A mental experience of joy or sorrow involves a relatively long duration, and there is no evidence whatsoever that such
mental experiences are faster than the time it takes to process the
body changes we have discussed. On the contrary, recent evidence
suggests that feelings occur over several seconds, two to twenty seconds being common.[23] Nonetheless, this objection has some merit
because if the system always operated precisely as James conceived,
it might not do the best job all of the time. I have proposed alterna-
tives that hinge on a critical notion: Feelings do not arise necessarily from the *actual body states*—although they can—but rather from
the *actual maps* constructed at any given moment in the body-
sensing regions. Against the background of these two issues we are
now ready to discuss my view of how the feeling system is organized and operates.

Actual Body States and Simulated Body States

At every moment of our lives the brain's body-sensing regions receive signals with which they can construct maps of the ongoing
body state. We can picture these maps as a set of correspondences
from everywhere and anywhere in the body toward the body-
sensing regions. This limpid picture of engineering clarity, how-

ever, is blurred by the fact that other brain regions can either interfere directly with the signaling toward the body-sensing regions, or
interfere directly with the activity of the body-sensing regions
themselves. The result of these "interferences" is most curious. As
far as our conscious mind is concerned there is only one source of
knowledge for what is going on in the body: the pattern of activity
present at any given moment in the body-sensing regions. Consequently, any interference with this mechanism can create a "false"
map of what is transpiring in the body at a particular moment.

NATURAL ANALGESIA

A good example of "false" body mapping occurs under certain circumstances when the brain filters out nociceptive body signals.
The brain effectively eliminates from the central body maps the
patterns of activity that would permit the experience of pain.
There are good reasons why the mechanisms of "false" representation would have prevailed in evolution. During an attempt to
run away from danger it is helpful not to feel the pain that may
come from wounds inflicted by the cause of danger (e.g., a bite
from a predator) or by the very act of fleeing from danger (running away and being hurt by obstacles).

We now have detailed evidence about how this sort of interference occurs. Nuclei in the part of the brain stem tegmentum
known as the periaqueductal gray (PAG) dispatch messages toward the nerve pathways that normally would convey signals of tissue damage and lead to the experience of pain. Those messages
prevent the signals from passing on.[24] Naturally, as a result of the
filtering, we get a "false" body map. The body-relatedness of the
process is not in question, of course. The dependence of feeling on
the "language" of body signals still is affirmed. It is just that what
we actually feel is not exactly what we would have felt without the
wise interference of the brain. This effect of the interference is

equivalent to taking a higher dose of aspirin or morphine, or being placed under local anesthesia. Except, of course, that the brain is doing it for you and it all comes naturally. Incidentally, the morphine metaphor applies quite closely because one of the varieties of this interference uses naturally and internally generated morphine analogues—opioid peptides such as the endorphins. There are several classes of opioid peptides, all of them naturally manufactured in our own body and thus called "endogenous." They include the endormorphines, enkephalin, and dynorphin, in addition to the endorphins. These molecules bind to specific classes of receptor in certain neurons of certain regions of the brain. Thus on some occasions of need, nature provides us with the same analgesic shot that the compassionate physician administers to the patient suffering pain.

We can find evidence for these mechanisms all around us. Those of us, public speakers or actors, who have had to perform while sick have experienced the strange disappearance of the worst physical symptoms of any ailment when we walk on a stage. Old wisdom credits the miracle change to the performer's "adrenalin rush." The notion that a chemical molecule is involved is wise indeed, but it does not tell us where the molecule acts and why the action causes the desired effect. I believe that what happens is a highly convenient modification of the current body maps. The modification requires several neural messages and does involve certain chemical molecules, although adrenalin probably is not the principal one. Soldiers in the battlefield also modify the body maps that portray pain and fear in their brains. Without that modification, acts of heroism would be less likely to occur. If this nice feature had not been added to the menu of our brains, evolution even might have discontinued childbirth in favor of a less painful variety of reproduction.

I suspect some notorious psychopathological conditions hijack this nice mechanism for good measure. The so-called hysterical or conversion reactions that allow patients not to feel or move parts of their body might well be consequent to transient but radical changes in current body maps. Several "somatoform" psychiatric disorders can be explained this way. Incidentally, a simple twist on these mechanisms might help suppress the recall of events that once caused marked anguish in our lives.

EMPATHY

It also is apparent that the brain can simulate certain emotional body states internally, as happens in the process of turning the emotion sympathy into a feeling of empathy. Think, for example, of being told about a horrible accident in which someone was badly injured. For a moment you may feel a twinge of pain that mirrors in your mind the pain of the person in question. You feel as if you were the victim, and the feeling may be more or less intense depending on the dimension of the accident or on your knowledge of the person involved. The presumed mechanism for producing this sort of feeling is a variety of what I have called the "as-if-body-loop" mechanism. It involves an internal brain simulation that consists of a rapid modification of ongoing body maps. This is achieved when certain brain regions, such as the prefrontal/premotor cortices, directly signal the body-sensing brain regions. The existence and location of comparable types of neurons has been established recently. Those neurons can represent, in an individual's brain, the movements that very brain sees in another individual, and produce signals toward sensorimotor structures so that the corresponding movements are either "previewed," in simulation mode, or actually executed. These neurons are present in the frontal cortex of monkeys and humans, and are known as "mirror neurons."[25] I believe the "as-if-body-loop" mechanism that I postulated in *Descartes' Error* draws on a variant of this mechanism.

The result of direct simulation of body states in body-sensing regions is no different from that of filtering of signals hailing from the body. In both cases the brain momentarily creates a set of body maps that does *not* correspond exactly to the current reality of the body. The brain uses the incoming body signals like clay to sculpt a particular body state in the regions where such a pattern can be constructed, i.e., the body-sensing regions. What one feels then is based on that "false" construction, not on the "real" body state.

A recent study from Ralph Adolph speaks directly to the issue of simulated body states.[26] The study was aimed at investigating the underpinnings of empathy and involved more than 100 patients with neurological lesions located at varied sites of their cerebral cortex. They were asked to participate in a task that called for the sort of process needed for empathy responses. Each subject was shown photographs of an unknown person exhibiting some emotional expression and the task consisted of indicating what the unknown person was feeling. Researchers asked each subject to place himself or herself in the person's shoes to guess the person's state of mind. The hypothesis being tested was that patients with damage to body-sensing regions of the cerebral cortex would not be capable of performing the task normally.

Most patients performed this task easily, precisely as healthy subjects do, except for two specific groups of patients whose performance was impaired. The first group of impaired patients was quite predictable. It was made up of patients with damage to visual association cortices, especially the right visual cortices of the ventral occipito-temporal region. This sector of the brain is critical for the appreciation of visual configurations. Without its integrity, the facial expressions in the photographs cannot be perceived as a whole, even if the photos can be seen in the general sense of the term.

The other group of patients was the most telling: It consisted of subjects with damage located in the overall region of the *right* somatosensory cortices, namely, in the insula, SII, and SI regions of the right cerebral hemisphere. This is the set of regions in which the brain accomplishes the highest level of integrated mapping of body state. In the absence of this region, it is not possible for the brain to simulate other body states effectively. The brain lacks the playground where variations on the body-state theme can be played.

It is of great physiological significance that the comparable region of the left cerebral hemisphere does not have the same function: Patients with damage to the left somatosensory complex perform the "empathy" task normally. This is one more finding that suggests that the right somatosensory cortices are "dominant" with regard to integrated body mapping. This is also the reason why damage to this region has been consistently associated with defects in emotion and feeling, and with conditions such as anosognosia and neglect, whose basis is a defective idea of the current body state.[27] The right versus left asymmetry in the function of the human somatosensory cortices probably is due to a committed participation of the left somatosensory cortices in language and speech.

Other supporting evidence comes from studies in which normal individuals who were viewing photographs depicting emotion immediately and subtly activated the muscular groups of their own faces that would have been necessary for them to make the emotional expressions depicted in the photographs. The individuals were not aware of this mirror-image "presetting" of their own muscles but electrodes distributed across their faces picked up on electromyographic changes.[28]

In summary, the body-sensing areas constitute a sort of theater where not only the "actual" body states can be "performed," but varied assortments of "false" body states can be enacted as well, for example, as-if-body states, filtered body states, and so on. The commands for producing as-if-body states are likely to come from a variety of prefrontal cortices as suggested by recent work on mirror-neurons in both animals and humans.

Hallucinating the Body

The brain allows us to *hallucinate* certain body states by a variety of means. One can imagine how such a feature began in evolution. At first the brain merely produced straight mappings of the body state. Later, other possibilities arose, for example, temporarily eliminating the mapping of body states such as those culminating in pain. Later perhaps, there was the possibility of simulating states of pain where none existed. These new possibilities clearly had their advantages and because those who had those advantages available prospered, the possibilities prevailed accordingly. As is the case with other valuable features of our natural makeup, pathological variations can corrupt the valuable use, as seems to be the case in hysteria and like disorders.

One additional practical value of these mechanisms is their speed. The brain can achieve the modification of body maps very rapidly in the time scale of hundreds of milliseconds or less, the brief period required by short and myelinated axons to convey signals from, say, the prefrontal cortex to the somatosensing maps of the insula, which lie just a few centimeters away. The time scale for the brain to induce changes in the body proper is seconds. It takes about one second for long and often unmyelinated axons to convey signals to body parts located tens of centimeters away from the brain. This is also the time scale for a hormone to be released into the bloodstream and begin to produce its cascade of subsequent effects. This is probably the reason why, in so many circumstances, we can sense an exquisite temporal relationship between subtle shades of feeling and the thoughts that prompted them or are consequent to them. The fast speed of as-if-body mechanisms brings thought and effected feeling close together in time, more easily so than if feeling were solely dependent on actual body changes.

It is worth noting that hallucinations such as we have been describing are not adaptive when they occur in sensory systems other than the one that has to do with the body's interior. Visual hallucinations are highly disruptive and so are auditory hallucinations. There is no benefit to them and they are not enjoyed as entertainment by the neurologic and psychiatric patients who have to suffer them. The same applies to the hallucinated smells or tastes that epileptic patients may experience. Yet body-state hallucinations, outside of the few psychopathological conditions I outlined, are valuable resources for the normal mind.

The Chemicals of Feeling

By now everyone knows that the so-called mood-altering drugs turn feelings of sadness or inadequacy into those of contentment and confidence. Long before the days of Prozac, however, alcohol, narcotics, analgesics, and hormones such as estrogens and testosterone, along with a host of psychotropic drugs, had shown that feelings can be altered by chemical substances. It is obvious that the action of all these chemical compounds is due to the design of their molecules. How do these compounds produce their noteworthy effects? The explanation usually is that chemical molecules act on certain neurons in certain brain regions to produce a desired result. From the standpoint of neurobiological mechanisms, however, these explanations sound a lot like magic. Tristan and Isolde drink the love potion; bang; and by the next scene they have fallen in love. It is not clear at all why having chemical X attach itself to the neurons of brain area Y can suspend your anguish and make you feel loving. What is the explanatory value of saying that male adolescents can become violent and hypersexed when they are awash with fresh testosterone? There is a functional level of explanation missing between the testosterone molecule and the adolescent behavior.

The incompleteness of the explanation comes from the fact that the actual origin of feeling states—their mental nature—is not conceptualized in neurobiological terms. The molecular level explanation is a part of solving the puzzle but does not get quite to what we really wish to see explained. The molecular mechanisms that result from the introduction of a drug in the system account for the beginning of the chain of processes that lead to the alteration of feeling but not for the processes that eventually establish the feeling. Little is said about what particular neural functions are altered by a drug so that feelings are altered. Little is said about which systems support those functions. We know the location of neuron receptors onto which certain chemical molecules may potentially attach themselves. (For example, we know that opioid receptors of the mu class are located in brain regions, such as the cingulate cortex, and we know that external as well as internal opioids act through attachment to those receptors.[29]) We know that the attachment of molecules to those receptors causes a change in the operation of the neurons equipped with those receptors. As a result of opioid binding to the mu-receptors of certain cortical neurons, neurons in the ventral tegmental area of the brain stem become active and lead to the release of dopamine in structures such as the nucleus accumbens of the basal forebrain. In turn, a number of rewarding behaviors occur, and a pleasurable feeling will be felt.[30] The neural patterns that form the basis for feelings, however, do not occur in the neurons of the aforementioned regions alone, and the actual "constitutive" patterns of feelings probably do not occur in those neurons at all. In all likelihood, the critical neural patterns, those that are the proximate cause of the feeling state, occur elsewhere—namely, in body-sensing regions such as the insula—as a result of the actions of the neurons directly affected by the chemical molecules.

Within the framework I have been constructing, we can specify processes that lead to altered feeling and we can specify loci for the action of drugs. If feelings arise from neural patterns that map myriad aspects of the ongoing body state, then the parsimonious hypothesis is that mood-altering chemicals produce their magic by changing the pattern of activity in those body-sensing maps. They can do so by means of three different mechanisms, working separately or in conjunction: One mechanism interferes with the transmission of signals from the body; another works by creating a particular pattern of activity within the body maps; yet another works by changing the very state of the body. All these mechanisms are open for drugs to perform their sleight of hand.

Varieties of Drug-Induced Felicity

Several lines of evidence indicate the importance of the brain's body-sensing maps as a basis for the generation of feelings. As noted, introspective analysis of normal feelings points unequivocally to the perception of varied body changes during the unfolding of feelings. The numerous functional imaging experiments discussed earlier reveal altered patterns of activity in body-sensing regions as a correlate of feelings. Another intriguing source of evidence is the introspective analysis of substance abusers who take drugs with the express purpose of producing an intense state of happiness. The first-person accounts of substance abusers contain frequent references to altered changes in the body during the drug highs. Here are some typical accounts:

My body was full of energy and at the same time completely relaxed.

It feels like every cell and bone in your body is jumping with delight.

There is a mild anesthetic property . . . and a generalized tingly, warm sensation.

It felt like a total body orgasm.

There is a pervasive body warmth.

The hot bath was so good I could not speak.

It felt like your head blowing up . . . a pleasant warmness and intense feeling of relaxation.

It's like the relaxed feeling you get after sex but better.

A body high.

A pins and needles effect . . . the body telling you it is completely numb.

You feel as if you've been wrapped in the most pleasing, warm, and comfortable blanket in the world.

My body felt instantly warm, especially my cheeks, which felt quite hot.[31]

All of these accounts report a remarkably uniform set of changes in the body—relaxation, warmth, numbness, anesthesia, analgesia, orgastic release, energy. Again, it makes no difference whether these changes actually occur in the body and are conveyed to somatosensing maps, or are directly concocted in these maps, or both. The sensations are accompanied by a set of syntonic thoughts—thoughts of positive events, an increased capacity "to understand," physical and intellectual power, removal of barriers and preoccupations. Curiously, the first four accounts came after cocaine highs. Ecstasy users reported the next three, and heroin users reported the last five. Alcohol produces more modest but comparable effects. The fact that the effects share a body core is all the more impressive considering that *the sub-*

stances that caused them are chemically different and act on different chemical systems in the brain. All of these substances act by occupying brain systems as if the molecules were being created from the inside. For example, cocaine and amphetamine act on the dopamine system. But the currently fashionable variant of amphetamine known as ecstasy (a mouthful of a molecule known as methylenedioxymethamphetamine or MDMA) acts on the serotonin system. As we have just seen, heroin and other opium-related substances act on the μ and δ opioid receptors. Alcohol works through GABA A receptors and through the NMDA glutamate receptors.[32]

It is important to note that the same systematic engagement of somatosensing regions described earlier in functional imaging studies of varied natural feelings can be found in studies in which the participants experienced feelings resulting from taking ecstasy, heroin, cocaine, and marijuana, or from craving such substances. Again, the cingulate cortex and the insula are the dominating sites of engagement.[33]

The anatomical distribution of the receptors on which these different substances act is quite varied as well, the pattern being somewhat different for each of the drugs. And yet the feelings they produce are quite similar. It is reasonable to propose that, in one way or another, at some point in their action, the different molecules help shape similar patterns of activity in the body-sensing regions. In other words, the feeling effect comes from changes in a shared neural site, or sites, which result from different cascades of system changes initiated by different substances. An account at the level of molecules and receptors alone is not sufficient to explain the effects.

Because all feelings contain some aspect of pain or pleasure as a necessary ingredient, and because the mental images we call feelings arise from the neural patterns exhibited in body maps, it is reasonable to propose that pain and its variants occur when the brain's body maps have certain configurations.

Likewise, pleasure and its variants are the result of certain map configurations. Feeling pain or feeling pleasure consists of having biological processes in which our body image, as depicted in the brain's body maps, is conformed in a certain pattern. Drugs such as morphine or aspirin alter that pattern. So do ecstasy and scotch. So do anesthetics. So do certain forms of meditation. So do thoughts of despair. So do thoughts of hope and salvation.

Enter the Naysayers

Some naysayers, while accepting the foregoing discussion on the physiological basis of feeling, will remain unsatisfied and claim that I still have not explained why feelings feel the way they do. I could reply that their question is ill-posed, that feelings feel the way they do because they just do, because that is just the nature of things. But I take their point and I have not run out of arguments. Let me continue, then, by adding detail to the answers given so far and by indicating with as much specificity as is possible the intimate nature of the mappings that contribute to a feeling.

At a glance, the body mappings underlying feeling may appear as a rough and vague representation of the state of viscera or muscles. But think again. Consider, first of all, that literally every region of the body is being mapped at the very same time because every region of the body contains nerve endings that can signal back to the central nervous system as to the state of the living cells that constitute that particular region. The signaling is complex. It is not a matter of "zeros" or "ones" indicating, for example, that a living cell is on or off. The signals are highly variegated. For example, nerve endings can indicate the magnitude of the concentration of oxygen and carbon dioxide in the vicinity of a cell. They can index the pH of the chemical bath in which every living cell is immersed. They can signal the presence of toxic compounds, external or internal. They also can detect the appearance of internally generated chemical molecules such as cytokines that indicate distress and impending disease for a living cell. In addition, nerve endings can indicate the state of contraction of muscle fibers, all the way from the smooth muscle fibers that constitute the wall of every artery, big or small, anywhere in the body, to the large, striated muscle fibers that constitute the muscles of our limbs, chest wall, or face. Nerve endings can thus indicate to the brain what viscera such as the skin or the gut are doing at any given moment. Moreover, in addition to the information they get from nerve endings, the body mappings that constitute the substrates of feelings in the brain are also directly informed about myriad variations in the concentration of chemical molecules in the bloodstream, via a nonneural route.

For example, in the part of the brain known as the hypothalamus, groups of neurons directly read the concentration of glucose (sugar) or water in your blood and take action accordingly. The action they take, as mentioned earlier, is designated as a drive or appetite. A diminishing concentration of glucose leads to the production of an appetite—the state of hunger—and to the initiation of behaviors aimed at ingesting food and eventually correcting the lowered glucose level. Likewise, a diminishing concentration of water molecules leads to thirst and water conservation. This is achieved by ordering the kidneys not to eliminate as much water and by changing the respiratory pattern so that less water is lost in the air we exhale. A number of other sites, namely the area postrema in the brain stem, and the subfornical organs, near the lateral ventricles, behave like the hypothalamus. They convert chemical signals conveyed by the bloodstream into neural signals transmitted along neural pathways inside the brain. The result is the same: The brain gets to map the state of the body.

As the brain surveys the entire organism, locally and directly—via nerve endings—and globally and chemically—via the bloodstream—the detail of these maps and their variegation are quite remarkable. They perform samplings of the state of life throughout the living organism, and from those amazingly extensive samplings they can distill integrated state maps. I suspect that when we say that we feel well or that we feel rotten, the sensation we experience is drawn from composite samplings based on the mapping of the internal milieu chemistries. It may be quite inaccurate to say, as we often do, that the neural signaling that goes on in the brain stem and hypothalamus is never conscious. I believe that a part of it is continually made conscious in a particular form and that it is precisely what constitutes our background feelings. It is true that background feelings may go unattended, but that is another matter; they are attended often enough. Think of this the next time you feel you are coming down with a cold, or, better still, that you are on top of the world and could not be any more fortunate than you are.

More Naysayers

More naysayers rise at this point to say that the cockpits of modern airplanes are filled with sensors for the body of the airplane much like the ones I am describing here. They ask me: Does the plane feel? And if so, do I know why it feels as it does?

Any attempt to associate what happens in a complex living organism to what happens in a splendidly engineered machine, say a Boeing 777, is foolhardy. It is true that on-board computers of a sophisticated aircraft include maps that monitor a variety of functions at any given moment: the state of deployment of the moving parts of the wings, of the horizontal stabilizer, and of the rudder; varied parameters in the operation of the engines; the consumption of fuel. Also monitored are ambient variables such as temperature, wind speed, altitude, and so forth. Some of the computers interrelate the monitored information continuously so that intelligent corrections can be made to the plane's ongoing behavior. The similarity to homeostatic mechanisms is obvious. But there are notable, nay, huge differences between the nature of the maps in the brain of a living organism and the cockpit of the Boeing 777. Let us consider them.

First, there is the scale of the detail with which the component structures and operations are represented. The monitoring devices in the cockpit are but a pale version of the monitoring devices in the central nervous system of a complex living organism. They are roughly comparable in our body to indicating whether our legs are crossed or uncrossed; to measuring heartbeat and body temperature; and to telling us how many hours we can go before eating the next meal. Very helpful but not quite enough for survival. Now, my point is not to put down the wonderful 777. My point is that the 777 does not need any more monitoring than it has to survive. Its "survival" is tied to living pilots who manage it and without whom the whole exercise is senseless. The same applies, incidentally, to the unmanned drones we fly around the world. Their "life" depends on mission control.

Some of the components of the aircraft are "animated"—slats and flaps, rudder, air brakes, undercarriage—but none of those components is "alive" in the biological sense. None of those components is made of cells whose integrity depends on the delivery of oxygen and nutrients to *each* of them. On the contrary, *every* elementary part of our organism, *every* cell in the body, is not just animated but living. Even more dramatically, every cell is an *individual* living organism—an individual creature with a birth date, life cycle, and likely death date. Each cell is a creature that must look after its own life and whose living is dependent upon the instructions of its own genome and the circumstances of its environment. The innate life-regulation devices that I discussed earlier in relation to humans are present down the biological scale in every system of our organism, in every organ, in every tissue, in every cell. The reasonable candidate for the title of critical elementary "particle" of our living organism is a living cell, not an atom.

There is nothing really equivalent to that living cell in the tons of aluminum, composite alloys, plastic, rubber, and silicone that make up the great Boeing bird. There are miles of electrical wiring, thousands of square feet of composite alloys, and millions of nuts, bolts, and rivets in the skin of the aircraft. It is true that all of these are made of matter, which is made of atoms. So is our human flesh at the level of its microstructure. But the physical matter of the aircraft is not alive, its parts are not made of living cells possessed of a genetic inheritance, a biologic destiny, and a life risk. And even if one were to argue that the plane has an "engineered concern" for its survival, which allows it to preempt the wrong maneuver of a distracted pilot, the blatant difference is inescapable. The plane's integrated cockpit computers have a concern for the execution of its flying function. Our brains and minds have a global concern for the integrity of our entire living real estate, every nook and cranny of it, and underneath it all, every nook and cranny has a local, automated concern with itself.

These distinctions are chronically glossed over whenever living organisms and intelligent machines, e.g., robots, are compared. Here I just wish to make clear that our brains receive signals from deep in the living flesh and thus provide local as well as global maps of the intimate anatomy and intimate functional state of that living flesh. This arrangement, so impressive in any complex living organism, is positively astounding in humans. I do not wish to diminish in any way the value of the interesting artificial creatures being created in the laboratories of Gerald Edel-

man or Rodney Brooks. In different ways, those engineered creatures deepen our understanding of certain brain processes and
may become useful complements of our own brain equipment. I
simply want to note that these animated creatures are not living in
the sense we are and are not likely to feel in the way we do.[34]

Notice something quite curious and also chronically over-
looked: The nerve sensors that convey the requisite information
to the brain and the nerve nuclei and nerve sheaths that map the
information inside of it *are living cells themselves, subject to the same
life risk of other cells, and in need of comparable homeostatic regulation.* These nerve cells are not impartial bystanders. They are not
innocent conveyances or blank slates or mirrors waiting for something to reflect. Signaling and mapping neurons have a say on the
matter signaled, and on the transient maps assembled from the
signals. The neural patterns that the body-sensing neurons assume hail from all the body activities they are meant to portray.
Body activities shape the pattern, give it a certain intensity and a
temporal profile, all of which contribute to why a feeling feels a
certain way. But in addition the *quality* of the feelings probably
hinges on the intimate design of the neurons themselves. The ex-
periential quality of the feeling is likely to depend on the medium
in which it is realized.

Finally, notice something quite intriguing and again neglected about the nature of the animation in the Boeing's moving
parts and in our living bodies. The Boeing's animation pertains
to the purposes of the functions the plane has been designed to
perform—taxi onto a runway, take off, fly, land. The equivalent in
our bodies is the animation that occurs when we look, listen,
walk, run, jump, or swim. But note how that part of human animation is merely the tip of the iceberg when I talk about emotions
and their underpinnings. The hidden part of the iceberg concerns
the animation whose purpose is solely the managing of the life

state in the parts and in the whole of our organism. It is precisely that part of the animation that constitutes the critical substrate for feelings. There is no equivalent for that part of the animation in current intelligent machines. My answer to the last naysayer is that the 777 is unable to feel anything like human feelings because, among many other reasons, it does not have an equivalent to our interior life to be managed, let alone portrayed.

The explanation of why feelings feel the way they do begins this way: Feelings are based on composite representations of the state of life in the process of being adjusted for survival in a state of optimal operations. The representations range from the myriad components of an organism to the level of the whole organism. The way feelings feel is tied to:

1. The intimate design of the life process in a multicellular organism with a complex brain.
2. The operation of the life process.
3. The corrective reactions that certain life states automatically engender, and the innate and acquired reactions that organisms engage given the presence, in their brain maps, of certain objects and situations.
4. The fact that when regulatory reactions are engaged due to internal or external causes, the flow of the life process is made either more efficient, unimpeded, and easier, or less so.
5. The nature of the neural medium in which all of these structures and processes are mapped.

On occasion I have been asked how these ideas might explain the "negativity" or "positivity" of feelings, implying that explaining the positive or negative signal of feelings cannot be explained.

But is it really so? The point made in item four above is that there are organism states in which the regulation of life processes becomes efficient, or even optimal, free-flowing and easy. This is a well-established physiological fact. It is not a hypothesis. The feelings that usually accompany such physiologically conducive states are deemed "positive," characterized not just by absence of pain but by varieties of pleasure. There also are organism states in which life processes struggle for balance and can even be chaotically out of control. The feelings that usually accompany such states are deemed "negative," characterized not just by absence of pleasure but by varieties of pain.

Perhaps we can say with some confidence that positive and negative feelings are determined by the state of life regulation. The signal is given by the closeness to, or departure from, those states that are most representative of optimal life regulation. Incidentally, the "intensity" of feelings also is likely to be related to the degree of corrections necessary in negative states, and to the degree to which positive states exceed the homeostatic set point in the optimal direction.

I suspect that the ultimate quality of feelings, a part of why feelings feel the way they feel, is conferred by the neural medium. But a substantial part of the answer to why they feel the way they do pertains to the fact that the life governance processes are either fluid or strained. That is simply their way of operating given the strange state we call life and the strange nature of organisms— Spinoza's *conatus*—that drives them to endeavor to preserve themselves, come what may, until life is suspended by aging, disease, or externally inflicted injury.

The fact that we, sentient and sophisticated creatures, call certain feelings positive and other feelings negative is directly related to the fluidity or strain of the life process. Fluid life states are naturally preferred by our *conatus*. We gravitate toward them. Strained life states are naturally avoided by our *conatus*. We stay away. We can sense these relationships, and we also can verify that in the trajectory of our lives fluid life states that feel positive come to be associated with events that we call good, while strained life states that feel negative come to be associated with evil.

This is the time to refine the formulation I proposed early in this chapter. The origin of feelings is the body in a certain number of its parts. But now we can go deeper and discover a finer origin underneath that level of description: the many cells that make those body parts and exist both as individual organisms with their own *conatus* and as cooperative members of the regimented society we call the human body, held together by the organism's own *conatus*.

The contents of feelings are the configurations of body state represented in somatosensing maps. But now we can add that the transient patterns of body state do change rapidly under the mutual, reverberative influences of brain and body during the unfolding of an occasion of feeling. Moreover, both the positive/negative valence of feelings and their intensity are aligned with the overall ease or difficulty with which life events are proceeding.

Finally, we can add that the living cells that constitute the somatosensing brain regions, as well as the neural pathways that transmit signals from body to brain, are not likely to be indifferent pieces of hardware. They probably make a critical contribution to the quality of the perceptions we call feelings.

This also is the time to bring back together what I have split apart. One reason why I distinguish emotion and feeling has to do with a research intention: In order to understand the entire set of affective phenomena, it is helpful to break components apart, study their operations, and discern how those components articulate in time. Once we gain the desired understanding, or some of it anyway, it is just as important to put the parts of the mechanism together again so that we can behold the functional whole they constitute.

Making whole returns us to Spinoza's claim that body and mind are parallel attributes of the same substance. We split them under the microscope of biology because we want to know how that single substance works, and how the body and mind aspects are generated within it. After investigating emotion and feeling in relative isolation we can, for a brief moment of quiet, roll them together again, as affects.

CHAPTER 4

Ever Since Feelings

Of Joy and Sorrow

Armed with a preliminary view of what feelings may be, it is time to ask what feelings may be *for*. In our attempt to answer this question it is perhaps helpful to begin with a reflection on how joy and sorrow, the two emblems of our affective life, are achieved and what they represent.

The events are initiated by the presentation of a suitable object—the emotionally competent stimulus. The processing of the stimulus, in the specific context in which it occurs, leads to the selection and execution of a preexisting program of emotion. In turn, the emotion leads to the construction of a particular set of neural maps of the organism to which signals from the body-proper contribute prominently. Maps of a certain configuration are the basis for the mental state we call joy and its variants, something like a score composed in the key of pleasure. Other maps are the basis for the mental state we call sorrow, which in Spinoza's broad definition encompasses negative states such as anguish, fear, guilt, and despair. These are scores composed in the key of pain.

The maps associated with joy signify states of equilibrium for the organism. Those states may be actually happening or as if they were happening. Joyous states signify optimal physiological coordination and smooth running of the operations of life. They not only are conducive to survival but to survival with well-being. The states of joy also are defined by a greater ease in the capacity to act.

We can agree with Spinoza when he said that joy (*laetitia* in his Latin text) was associated with a transition of the organism to a state of greater perfection.[1] That is greater perfection in the sense of greater functional harmony, no doubt, and greater perfection in the sense that the power and freedom to act are increased.[2] But we should be mindful of the fact that the maps of joy can be falsified by a host of drugs and thus fail to reflect the actual state of the organism. Some of the "drug" maps may reflect a transient improvement of organism functions. Ultimately, however, the improvement is biologically untenable and is a prelude to a worsening of function.

The maps related to sorrow, in both the broad and narrow senses of the word, are associated with states of functional disequilibrium. The ease of action is reduced. There is pain of some kind, signs of disease or signs of physiological discord—all of which are indicative of a less than optimal coordination of life functions. If unchecked, the situation is conducive to disease and death.

In most circumstances the body maps of sorrow probably are reflective of the actual organism state. There are no drugs of abuse *meant* to induce sorrow and depression. Who would wish to take them, let alone abuse them? But the drugs of abuse do induce sorrow and depression on the rebound from the joyous highs they produce at first. For example, it is reported that the drug ecstasy produces highs characterized by a quietly pleasurable state and benign accompanying thoughts. Repeated use of the drug, however, induces more and more severe depressions, which follow highs that become less and less so. The normal operation of the serotonin system appears to be directly affected, and a drug many users consider safe proves to be quite dangerous.

In keeping with Spinoza when he discussed *tristitia*, the maps of sorrow are associated with the transition of the organism to a state of lesser perfection. The power and freedom to act are diminished. In the Spinozian view, the person in the throes of sadness is cut off from his or her *conatus*, from the tendency for self-preservation. This certainly applies to the feelings reported in severe depression, and to their ultimate consequences in suicide. Depression can be seen as part of a "sickness syndrome." The endocrine and immunological systems participate in sustained depression, as if a pathogen such as a bacterium or virus invaded the organism, destined to cause disease.[3] In isolation, occasions of sadness, fear, or anger are not likely to engage depression's downward spiral of sickness. The fact remains, however, that each and every occasion of negative emotion and subsequent negative feeling places the organism in a state outside its regular range of operations. When the emotion is fear, the special state may be advantageous—provided the fear is justified and not the result of an incorrect assessment of the situation or the symptom of a phobia. Justified fear is an excellent insurance policy, of course. It has saved or bettered many lives. But the engagement of anger or sadness is less helpful, personally and socially. Sure enough, well-targeted anger can discourage abuse of many sorts and act as a defensive weapon as it still does in the wild. In many social and political situations, however, anger is a good example of an emotion whose homeostatic value is in decline. The same could be said for sadness, a form of crying out for comfort and support with few tears. Still, sadness can be protective in the right circumstances, for example, when it helps us adapt to personal loss. In the long run, however, it is cumulatively harmful and can cause cancer, in this case, of the soul.

Feelings then can be mental sensors of the organism's interior, witnesses of life on the fly. They can be our sentinels as well. They let our fleeting and narrow conscious self know about the current state of life in the organism for a brief period. Feelings are the mental manifestations of balance and harmony, of disharmony and discord. They do not refer to the harmony or discord of objects or events out in the world, necessarily, but rather to the harmony or discord deep in the flesh. Joy and sorrow and other feelings are largely ideas of the body in the process of maneuvering itself into states of optimal survival. Joy and sorrow are mental revelations of the state of the life process, except when drugs or depression corrupt the fidelity of the revelation (although it could be argued that the sickness revealed by depression is, after all, faithful to the true state of life).

How intriguing that feelings bear witness to the state of life deep within. When we try to reverse the engineering of evolution and discover the origins of feelings, it is legitimate to wonder if bearing witness to life within our minds is the reason feelings prevailed as a prominent feature of complex living beings.

Feelings and Social Behavior

There is growing evidence that feelings, along with the appetites and emotions that most often cause them, play a decisive role in social behavior. In a number of studies published over the past two decades, our research team and others have shown that when previously normal individuals sustain damage to brain regions necessary for the deployment of certain classes of emotions and feelings, their ability to govern their lives in society is extremely disturbed. Their ability to make appropriate decisions is compromised in situations in which the outcomes are uncertain, such as making a financial investment or entering an important relationship.[4] Social contracts break down. More often than not, marriages dissolve, relations between parents and children strain, and jobs are lost.

After the onset of their brain lesion these patients are generally not able to hold on to their premorbid social status, and all of them cease to be financially independent. They usually do not become violent, and their misbehavior does not tend to violate the law. Nonetheless, the proper governance of their lives is profoundly affected. It is apparent that, if left to their own devices, their survival with well-being would be in serious question.

The typical patient with this condition was a hardworking and successful individual who performed a skilled job and earned a good living until the onset of the disease. Several patients we studied were active in social affairs and were even perceived by others as community leaders. After the onset of prefrontal damage, a completely changed person emerges. The patients remained skilled enough to hold a job but could not be counted on to report to work reliably or to execute all the tasks necessary for a goal to be accomplished. The ability to plan activities was impaired on a daily basis as well as in the long-term. Financial planning was especially compromised.

Social behavior is an area of particular difficulty. It is not easy for these patients to determine who is trustworthy and guide future behavior accordingly. Patients lack a sense of what is socially appropriate. They disregard social conventions and may violate ethical rules.

Their spouses note a lack of empathy. The wife of one of our patients noted how her husband, who previously reacted with care and affection anytime she was upset, now reacted with indifference in the same circumstances. Patients who prior to their disease were known to be concerned with social projects in their communities or who were known for their ability to counsel friends and relatives in difficulty no longer show any inclination to help. For practical purposes, they are no longer independent human beings.

When we ask ourselves why this tragic situation occurs we find a number of intriguing answers. The immediate cause of the problem is brain damage in a specific region. In the most serious and telling cases, those in which the disturbances of social behavior dominate the clinical picture, there is damage to some regions of the frontal lobe. The prefrontal sector, especially the part known as ventromedial, is involved in most, though not all, such cases. Damage restricted to the left lateral sectors of the frontal lobe tends not to cause this problem, although I know of at least one exception; damage restricted to the right lateral sector can.[5] [See figure 4.1.] Damage to a few other brain regions, namely the parietal sector of the right cerebral hemisphere, causes a similar

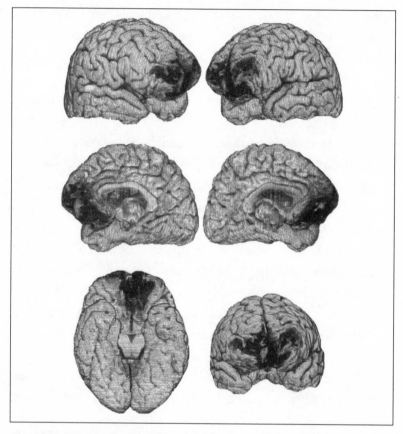

Figure 4.1: Pattern of prefrontal damage in a living adult patient shown in a three-dimensional reconstruction of his brain's magnetic resonance scan. The damage shows up in black and is easily distinguishable from the remainder of the intact brain. The top two panels show the brain seen from right and left hemisphere perspectives. The middle two panels show the medial (internal) views of the right and left cerebral hemispheres (respectively left middle panel and right middle panel). The bottom panels show the lesion seen from below (left) revealing the extensive damage to the orbital surface of the frontal lobe; and seen from the front (right), revealing the extensive damage of the frontal pole.

problem, although less pure, in the sense that other prominent neurological symptoms also are present. The parietal patients with comparable problems are usually paralyzed on the left side of their bodies, at least in part. The distinction of the patients with damage to the ventromedial sector of the frontal lobe is that their problems seem confined to their strange social behavior. For all intents and purposes they look normal.

The behavior of these prefrontal patients, however, is a world apart from how they acted prior to the neurological condition. They make decisions that are not advantageous for themselves and for those close to them. Yet, the patients appear to be intellectually intact. They talk normally, move normally, and do not have problems of visual or auditory perception. They are not distractible when engaged in a conversation. They learn and recall the facts that occur to them, they remember the conventions and rules they break every day, and they even realize, when someone calls their attention, that they have broken those conventions and rules. They are intelligent in the technical sense of the term, that is, they can score highly on IQ measurements. They can solve logical problems.

For a long time, attempts were made to account for these patients' poor decision-making on the basis of cognitive failures. Perhaps their problem was one of learning or recalling the material necessary to behave properly. Perhaps their problem was one of reasoning intelligently through the material. Or perhaps the difficulty was something as simple as holding in mind, for the necessary period of time, all the premises of the problem that needed to be taken into consideration for a proper solution (this "holding-in-mind" function is known as "working memory").[6] None of these explanations was satisfactory, however. Somehow, the majority of these patients do not have a primary problem in any of those presumably impaired capacities. It is quite disconcerting to hear one of those patients reason intelligently and solve successfully a specific social problem when the problem is presented in the laboratory, as a test, in the form of a hypothetical situation. The problem may be precisely the same kind the patient has just failed to solve in real life and real time. These patients exhibit extensive knowledge about the social situations that they so egregiously mismanaged in reality. They know the premises of the problem, the options of action, the likely consequences of those actions immediately and in the long-term, and how to navigate such knowledge logically.[7] But all of this is to no avail when they need it most in the real world.

Inside a Decision-Making Mechanism

While studying these patients I became intrigued by the possibility that their reasoning defect was tied not to a primarily cognitive problem, but rather to a defect in emotion and feeling. Two factors contributed to this hypothesis. First, there was the obvious failure of accounting for the problem on the basis of the more obvious cognitive functions. Second, and most important, I had become aware of the degree to which such patients are emotionally flat at the level of their social emotions. I was especially struck by the fact that emotions such as embarrassment, sympathy, and guilt appear diminished or absent. I felt sadder and more embarrassed over the personal stories some of the patients told me than they themselves seemed to be.[8]

This is how I came to the idea that the reasoning defect these patients exhibited, their defect in the governance of life, might be due to the impairment of an emotion-related signal. I was suggesting that when these patients faced a given situation—its options for action, and the mental representation of the outcomes of the possible actions—they failed to activate an emotion-related memory that would have helped them choose more advantageously among competing options. The patients were not making use of the emotion-related experience they had accumulated in their lifetimes. Decisions made in these emotion-impoverished circumstances led to erratic or downright negative results, especially so in terms of future consequences. The compromise was most notable for situations involving markedly conflicting options and uncertainty of outcomes. Choosing a career, deciding whether to marry, or launching a new business are examples of decisions whose outcomes are uncertain, regardless of how carefully prepared one may be when the decision is made. Typically one has to choose among conflicting options, and emotions and feelings come in handy in those circumstances.

How could emotion and feeling play a role in decision-making? The answer is that there are many ways, subtle and not so subtle, practical and not so practical, all of which make emotion and feeling not merely players in the process of reasoning, but indispensable players. Consider that, for example, as personal experience is accumulated, varied categories of social situation are formed. The knowledge we store regarding those life experiences includes:

1. The facts of the problem presented;
2. The option chosen to solve it;
3. The factual outcome to the solution, and, importantly,
4. The outcome of the solution in terms of emotion and feeling.

For example, did the immediate outcome of the chosen action bring punishment or reward? In other words, was it accompanied by emotions and feelings of pain or pleasure, sorrow or joy, shame or pride? No less importantly, was the *future outcome* of the actions punishing or rewarding, regardless of how positive or negative the immediate outcome? How did things work out in the long run? Were there negative or positive future consequences resulting from the specific action? In a typical instance, did break-ing or starting a certain relationship lead to benefits or disaster?

The emphasis on future outcomes calls attention to something quite particular about human behavior. One of the main traits of civilized human behavior is thinking in terms of the future. Our baggage of accumulated knowledge and our ability to compare past and present have opened the possibility of "minding" the future, predicting it, anticipating it in simulated form, attempting to shape it in as beneficial a manner as possible. We trade instantaneous gratification and defer immediate pleasure for a better future, and we make immediate sacrifices on the same basis.

As we argued earlier, every experience in our lives is accom-panied by some degree of emotion and this is especially obvious in relation to important social and personal problems. Whether the emotion is a response to an evolutionarily set stimulus, as is often the case in sympathy, or to a learned stimulus, as in the case of apprehension acquired by association with a primary fear stimulus, it does not matter: positive or negative emotions and their ensuing feelings become obligate components of our social experiences.

The idea then is that, over time, we do far more than merely respond automatically to components of a social situation with the repertoire of innate social emotions. Under the influence of social emotions (from sympathy and shame, to pride and indig-nation) and of those emotions that are induced by punishment and reward (variants of sorrow and joy), we gradually categorize the situations we experience—the structure of the scenarios, their components, their significance in terms of our personal nar-rative. Moreover, we connect the conceptual categories we form—mentally and at the related neural level—with the brain apparatus used for the triggering of emotions. For example, different op-tions for action and different future outcomes become associated with different emotions/feelings. By virtue of those associations, when a situation that fits the profile of a certain category is revisited in our experience, we rapidly and automatically deploy the appropriate emotions.

In neural terms, the mechanism works like this: When circuits in posterior sensory cortices and in temporal and parietal regions process a situation that belongs to a given conceptual category, the prefrontal circuits that hold records pertinent to that category of events become active. Next comes activation of regions that trigger appropriate emotional signals, such as the ventromedial prefrontal cortices, courtesy of an acquired link between that category of event and past emotional-feeling responses. This arrangement allows us to connect categories of social knowledge—whether acquired or refined through individual experience—with the innate, gene-given apparatus of social emotions and their subsequent feelings. Among these emotions/feelings, I accord special importance to those that are associated with the future outcome of actions, because they come to signal a prediction of the future, an anticipation of the consequence of actions. This is a good example, incidentally, of how nature's juxtapositions generate complexity, of how putting together the right parts produces more than their mere sum. Emotions and feelings have no crystal ball to see the future. Deployed in the right context, however, they become harbingers of what may be good or bad in the near or distant future. The deployment of such anticipatory emotions/feelings can be partial or complete, overt or covert.

What the Mechanism Accomplishes

The revival of the emotional signal accomplishes a number of important tasks. Covertly or overtly, it focuses attention on certain aspects of the problem and thus enhances the quality of reasoning over it. When the signal is overt it produces automated alarm signals relative to options of action that are likely to lead to negative outcomes. A gut feeling can suggest that you refrain from a choice that, in the past, has led to negative consequences, and it can do so ahead of your own regular reasoning telling you precisely the same "Do not." The emotional signal can also produce the contrary of an alarm signal, and urge the rapid endorsement of a certain option because, in the system's history, it has been associated with a positive outcome. In brief, the signal *marks* options and outcomes with a positive or negative signal that narrows the decision-making space and increases the probability that the action will conform to past experience. Because the signals are, in one way or another, body-related, I began referring to this set of ideas as "the somatic-marker hypothesis."

The emotional signal is not a substitute for proper reasoning. It has an auxiliary role, increasing the efficiency of the reasoning process and making it speedier. On occasion it may make the reasoning process almost superfluous, such as when we immediately reject an option that would lead to certain disaster, or, on the contrary, we jump to a good opportunity based on a high probability of success.

In some cases the emotional signal can be quite strong, leading to a partial reactivation of an emotion such as fear or happiness, followed by the appropriate conscious feeling of that emotion. This is the presumed mechanism for a gut feeling, which uses what I have called a body-loop. There are, however, subtler ways for the emotional signal to operate and presumably that is how emotional signals do their job most of the time. First, it is possible to produce gut feelings without actually using the body, drawing instead on the as-if-body-loop that I discussed in the previous chapter. Second, and more importantly, the emotional signal can operate entirely under the radar of consciousness. It can produce alterations in working memory, attention, and reasoning so that the decision-making process is biased toward selecting the action most likely to lead to the best possible outcome, given prior experience. The individual may not ever be cognizant of this covert operation. In these conditions we intuit a decision and enact it, speedily and efficiently, without any knowledge of the intermediate steps.

Our research team and others have accumulated substantial evidence in support of such mechanisms.9 The body-relatedness

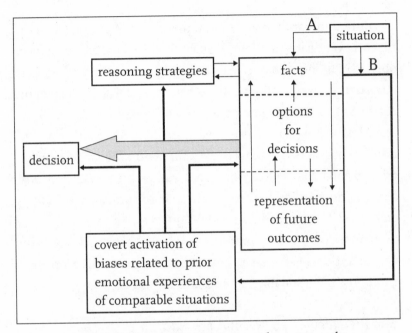

Figure 4.2: Normal decision-making uses two complementary paths. Confronted with a situation that requires a response, path A prompts images related to the situation, the options for action, and the anticipation of future outcomes. Reasoning strategies can operate on that knowledge to produce a decision. Path B operates in parallel and prompts activation of prior emotional experiences in comparable situations. In turn, the recall of the emotionally related material, be it covert or overt, influences the decision-making process by forcing attention on the representation of future outcomes or interfering with reasoning strategies. On occasion, path B can lead to a decision directly, as when a gut feeling impels an immediate response. The degree to which each path is used alone or in combination depends on a person's individual development, the nature of the situation, and the circumstances. The intriguing decision patterns described by Daniel Kahnemann and Amos Tversky in the 1970s are probably due to engagement of path B.

of the operation has been noted in the wisdom of the ages. The hunches that steer our behavior in the proper direction are often referred to as the gut or the heart—as in "I know in my heart that this is the right thing to do." The Portuguese word for hunch, by the way, is *palpite,* a close neighbor of "palpitation," a skipped heartbeat.

Although hardly mainstream, the idea that emotions are inherently rational has a long history. Both Aristotle and Spinoza obviously thought that at least some emotions, in the right circumstances, were rational. In a way, so did David Hume and Adam Smith. The contemporary philosophers Ronald de Sousa and Martha Nussbaum also have argued persuasively for the rationality of emotion. In this context the term rational does not denote explicit logical reasoning but rather an association with actions or outcomes that are beneficial to the organism exhibiting emotions. The recalled emotional signals are not rational in and of themselves, but they promote outcomes that could have been derived rationally. Perhaps a better term to denote this property of emotions is "reasonable," as suggested by Stefan Heck.[10]

The Breakdown of a Normal Mechanism

How does brain damage in previously normal adults bring on the defects of social behavior we described earlier? The damage causes two complementary impairments. It destroys the emotion-triggering region, where commands for the deployment of social emotions usually arise; *and* destroys the nearby region that supports the acquired link between certain categories of situation and the emotion that serves as the best guide to action in terms of future consequences. The repertoire of automated social emotions we have inherited cannot be deployed in response to the naturally competent stimuli, and neither can the emotions that we have learned to connect with certain situations in the course of individual experience. Furthermore, the subsequent feelings arising from all these emotions are also compromised. The severity of the defect varies from patient to patient. In every case, however, the patient becomes unable to produce in a reliable manner emotions and feelings tuned to specific categories of social situations.

The use of cooperative strategies of behavior appears to be blocked in patients with damage to brain regions such as the ventromedial frontal lobe. They fail to express social emotions and their behavior no longer observes the social contract. Their performance on tasks that depend on a deployment of social wisdom is abnormal.[11] Moreover, the use of cooperative strategies in normal individuals engages the ventromedial frontal regions as shown in functional imaging studies in which the participants were asked to solve the Prisoner's Dilemma, an experimental task that effectively separates cooperators from defectors. In a recent study, cooperativity also led to the activation of regions involved in the release of dopamine and in pleasure behavior, suggesting, well, that virtue is its own reward.[12]

Considering the condition of our adult-onset patients, one might have been tempted to predict that all their intact "social knowledge" and all that nice practice of social problem-solving prior to the onset of brain damage would have been sufficient to ensure normal social behavior. But this is simply not true. In one way or another, the factual knowledge about social behavior requires the machinery of emotion and feeling to express itself normally.

The myopia of the future caused by prefrontal damage has a counterpart in the condition of anyone who consistently alters normal feelings by taking narcotics or large quantities of alcohol. The resulting maps of life are systematically false, consistently misinforming brain and mind about the actual body state. One might guess that this distortion would be an advantage. What's wrong with feeling fine and being happy? Well, there seems to be a lot wrong, actually, if the well-being and happiness are substantially and chronically at variance with what the body would normally be reporting to the brain. In effect, in the circumstances of addiction, the processes of decision-making fail miserably and addicts progressively make less and less advantageous decisions for themselves and for those who are close to them. The term "myopia of the future" describes this predicament accurately. If left unchecked, it invariably leads to a loss of social independence.

It might be argued that in the case of addiction the impairments of decision could be due to the direct action of the drugs on neural systems supporting cognition in general rather than feeling in particular, but the explanation would be rather generous. Without the proper help, the well-being of addicts vanishes almost completely, except for the periods during which the abused substances create shorter and shorter occasions of pleasure. I suspect that the downward spiral of addicts' lives begins as a result of the distortions of feeling and the ensuing decision impairments, although eventually the physical ailments produced by chronic drug intake bring on further disease problems and often death.

Damage to Prefrontal Cortex in the Very Young

The findings and interpretations regarding the adult frontal lobe patients become especially compelling in light of the recent description of young adults, barely in their twenties, who sustained comparable frontal lobe damage early in life rather than in adulthood.[13] My colleagues Steven Anderson and Hanna Damasio are finding that those patients are, in many ways, similar to those who sustain lesions as adults. Just as in the adult cases, they do not exhibit sympathy, embarrassment, or guilt, and seem to have lacked those emotions and the corresponding feelings for their entire existence. But there are remarkable differences as well. The patients whose brain damage occurred during the first years of their lives have an even more severe defect in social behavior; more importantly, they seem never to have learned the conventions and rules that they violate. An example may help.

The very first patient we studied with this condition was twenty when we met her. Her family was comfortable and stable, and her parents had no history of neurological or psychiatric disease. She sustained head injury when she was run over by a car at age fifteen months, but she had recovered fully within days. No behavioral abnormalities were observed until age three when her parents noted she was unresponsive to verbal and physical punishment. This differed remarkably from the behavior of her siblings who went on to become normal adolescents and young adults. By fourteen her behavior was so disruptive that her parents placed her in a treatment facility, the first of many. She was academically capable yet routinely failed to complete her assignments. Her adolescence was marked by failure to comply with rules of any sort and frequent confrontations with peers and adults. She was verbally and physically abusive to others. She lied chronically. She was arrested several times for shoplifting and stole from other children and from her own family. She engaged in early and risky sexual behavior and became pregnant at eighteen. After the baby was born, her maternal behavior was marked by insensitivity to the child's needs. She was unable to hold any job due to poor dependability and violation of work rules. She never expressed guilt or remorse for inappropriate behavior or any sympathy for others. She always blamed others for her difficulties. Behavioral management and psychotropic medication were of no help. After repeatedly placing herself at physical and financial risk, she became dependent on her parents and on social agencies for both financial support and oversight of her personal affairs. She had no plans for the future and no desire to become employed.

This young woman had never been diagnosed with brain damage. Her history of early head injury had been virtually forgotten. Eventually, her parents wondered about a possible link and came to us. When we obtained a magnetic resonance scan of her brain, we found, as we expected, brain damage comparable to that of adult prefrontal patients. We have studied similar patients, all of whom showed the same pairing of abnormal social behavior and prefrontal damage. Our team is developing rehabilitation programs for patients such as these.

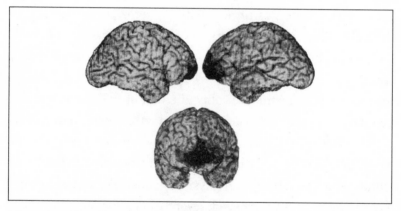

Figure 4.3: Three-dimensional reconstruction of the brain of a young adult who sustained damage to the prefrontal region early in life. As in the case shown in figure 4.1, the reconstruction is based on magnetic resonance data. Note the similarity to the area of damage seen in the adult patient.

We are not suggesting that every adolescent with similar behavior has undiagnosed brain damage. It is likely, however, that many people with comparable behavior that is not due to the same cause have a malfunction of the brain system that had sustained damage in our patients. The malfunction may be due to a defect in the operation of neural circuits at microscopic level. Such a defect may have a variety of causes, from abnormal chemical signaling on a genetic basis to social and educational factors.

Given the cognitive and neural setup we discussed earlier, we can understand why sustaining damage to the prefrontal region early in life has devastating consequences. The first consequence is that the innate social emotions and feelings are not deployed normally. At the very least, this leads the young patients to abnormal interactions with others. They react inappropriately in a host of social situations, and, in turn, others will react inappropriately to them. The young patients develop a skewed concept of the social world. Second, the young patients fail to acquire a repertoire of emotional reactions tuned to specific prior actions. This is because the learning of a connection between a particular action and its emotional consequences depends on the integrity of the prefrontal region. The experience of pain, which is part of punishment, becomes disconnected from the action that caused the punishment, and thus *there will not be a memory of their conjunction for future use;* likewise for the pleasurable aspects of reward. Third, there is a deficient individual buildup of personal knowledge about the social world. The categorizing of situations, the categorizing of adequate and inadequate responses, and the lay down and connection of conventions and rules, are distorted.[14]

What If the World?

There is little question that the integrity of emotion and feeling is necessary for normal human social behavior, by which I mean social behavior that conforms to ethical rules and laws and can be described as just. One shudders to think what the world would look like, socially speaking, if anything but a small minority of the population suffered from the condition visited upon humans with adult onset frontal lobe damage.

The shudder is even greater when we imagine a large population of patients who sustained damage to the frontal lobe early in their lives. It would be bad enough if such patients were rampant today. But one wonders how the world would have evolved if humanity had dawned with a population deprived of the ability to respond toward others with sympathy, attachment, embarrassment, and other social emotions that are known to be present in simple form in some nonhuman species.

EMBARRASSMENT; SHAME; GUILT
ECS: weakness/failure/violation in the individual's own person or
    behavior
consequences: prevent punishment by others (including ostracism,
    ridicule); restore balance in self, other, or group; enforcing of
    social conventions and rules
basis: fear; sadness; submissive tendencies

CONTEMPT; INDIGNATION
ECS: another individual's violation of norms (purity; cooperation)
consequences: punishment of violation; enforcing of social
    conventions and rules
basis: disgust; anger

SYMPATHY/COMPASSION
ECS: another individual in suffering/need
consequences: comfort, restoration of balance in other or group
basis: attachment; sadness

AWE/WONDER; ELEVATION; GRATITUDE; PRIDE
ECS: recognition (in others or in self) of a contribution to
    cooperation
consequences: reward for cooperation; reinforcing of tendency
    toward cooperation
basis: happiness

Figure 4.4: Some of the main social emotions both positive and negative. Under each group of emotions we identify the emotionally competent stimulus (ECS) capable of triggering the emotion; the main consequences of the emotion; and the physiological basis of the emotion. For more on the social emotions see text and the work of J. Haidt and R. Shweder.[15]

One might be tempted to dismiss this thought experiment summarily by saying that such a species would have been extinct soon. But please do not dismiss it so fast because that is precisely the point. In a society deprived of such emotions and feelings, there would have been no spontaneous exhibition of the innate social responses that foreshadow a simple ethical system—no budding altruism, no kindness when kindness is due, no censure when censure is appropriate, no automatic sense of one's own failings. In the absence of the feelings of such emotions, humans would not have engaged in a negotiation aimed at finding solutions for problems faced by the group, e.g., identification and sharing of food resources, defense against threats or disputes among its members. There would not have been a gradual build-up of wisdom regarding the relationships among social situations, natural responses, and a host of contingencies such as the punishment or reward incurred by permitting or inhibiting natural responses. The codification of rules eventually expressed in systems of justice and sociopolitical organizations is hardly conceivable in those circumstances, even assuming that the apparatus of learning, imagination, and reasoning could be otherwise intact in the face of the emotional ravages, a most unlikely possibility. With the natural system of emotional navigation more or less disabled, there would not have been a ready possibility of fine-tuning the individual to the real world. Moreover, the possibility of constructing a fact-based social navigation system, independently of the missing natural system, appears unlikely.

The dire scenario would apply equally well regardless of how one conceptualizes the origin of the ethical principles that guide social life. For example, if ethical principles emerged from a process of cultural negotiations conducted under the influence of social emotions, humans with prefrontal damage would not have engaged in such a process and would not have even begun constructing an ethical code. But the problem remains if one believes those principles came by way of religious prophecy handed over to a number of chosen humans. In one option, that of religion being one of the most extraordinary human creations, it is unlikely that humans without basic social emotions and feelings would ever have created a religious system in the first place. As we shall discuss in chapter 7, religious narratives may have arisen in response to important pressures, namely consciously analyzed sorrow and joy and the need to create an authority capable of validating and enforcing ethical rules. In the absence of normal emotions, there might have been no drive toward the creation of religion. There would have been no prophets, nor would there have been followers animated by the emotional tendency to submit with awe and admiration to a dominant figure entrusted with a leadership role, or to an entity with the power to protect and compensate for losses and the ability to explain the unexplainable. The conception of God, applied to one or many, would have been hard to come by.

Things would not fare better, however, if religious prophecies are presumed to have a supernatural origin, the prophet being a mere vehicle for the revealed wisdom. The ethical principles still would need to be inculcated in the developing and innocent human child with the leverage of punishment and reward, something that would be precluded in the situation of early onset prefrontal damage. Joy and sorrow, to the degree that they might be experienced in some circumstances by such individuals, would not connect with the categories of personal and social knowledge that define the fundamental issues of ethics. In brief, whether one sees ethical principles as mostly nature-based or religion-based developments, it appears that the compromise of emotion and feeling early in human development would not have boded well for the emergence of ethical behavior.

The elimination of emotion and feeling from the human picture entails an impoverishment of the subsequent organization of experience. If social emotions and feelings are not properly deployed, and if the relation between social situations and joy and sorrow breaks down, the individual cannot categorize the experience of events in his autobiographical memory record according to the emotion/feeling mark that confers "goodness" or "badness" upon those experiences. That would preclude any subsequent level of construction of the notions of goodness and badness, namely the reasoned cultural construction of what *ought* to be considered good or bad, given its good or bad effects.

Neurobiology and Ethical Behaviors

I suspect that in the absence of social emotions and subsequent feelings, even on the unlikely assumption that other intellectual abilities could remain intact, the cultural instruments we know as ethical behaviors, religious beliefs, laws, justice, and political organization either would not have emerged, or would have been a very different sort of intelligent construction. A word of caution, however. I do not mean to say that emotions and feelings singlehandedly caused the emergence of those cultural instruments. First, the neurobiological dispositions likely to facilitate the emergence of such cultural instruments include not just emotions and feelings, but also the capacious personal memory that allows humans to construct a complex autobiography, as well as the process of extended consciousness that permits close interrelations among feelings, self, and external events. Second, a simple neurobiological explanation for the rise of ethics, religion, law, and justice is hardly viable. It is reasonable to venture that neurobiology will play an important role in future explanations. But in order to comprehend these cultural phenomena satisfactorily we need to factor in ideas from anthropology, sociology, psychoanalysis, and evolutionary psychology, as well as findings from studies in the fields of ethics, law, and religion. In fact, the course most likely to yield interesting explanations is a new breed of investigations aimed at testing hypotheses based on integrated knowledge from any and all of these disciplines *and* neurobiology.[16] Such an undertaking is barely beginning to take shape and, at any rate, goes beyond the scope of this chapter and of my preparation. It does seem sensible, however, to suggest that feelings may have been a necessary grounding for ethical behaviors long before the time humans even began the deliberate construction of intelligent norms of social conduct. Feelings would have entered the picture in prior evolutionary stages of nonhuman species, and would have been a factor in the establishment of automated social emotions and of cognitive strategies of cooperativity. My position on the intersection of neurobiology and ethical behavior can be summarized with the following statements.

Ethical behaviors are a subset of social behaviors. They can be investigated with a full range of scientific approaches, from anthropology to neurobiology. The latter encompasses techniques as diverse as experimental neuropsychology (at the level of large-scale systems) and genetics (at the molecular level). The most fruitful results are likely to come from combined approaches.[17]

The essence of ethical behavior does not begin with humans. Evidence from birds (such as ravens) and mammals (such as vampire bats, wolves, baboons, and chimpanzees) indicates that other species can behave in what appears, to our sophisticated eyes, as an ethical manner. They exhibit sympathy, attachments, embarrassment, dominant pride, and humble submission. They can censure and recompense certain actions of others. Vampire bats, for example, can detect cheaters among the food gatherers in their group and punish them accordingly. Ravens can do likewise.

Such examples are especially convincing among primates, and are by no means confined to our nearest cousins, the big apes. Rhesus monkeys can behave in a seemingly altruistic manner toward other monkeys. In an intriguing experiment conducted by Robert Miller and discussed by Marc Hauser, monkeys abstained from pulling a chain that would deliver food to them if pulling the chain also caused another monkey to receive an electric shock. Some monkeys would not eat for hours, even days. Suggestively, the animals most likely to behave in an altruistic manner were those that knew the potential target of the shock. Here was compassion working better with those who are familiar than with strangers. The animals that previously had been shocked also were more likely to behave altruistically. Nonhumans can certainly cooperate or fail to do so, within their group.[18] This may displease those who believe just behavior is an exclusively human trait. As if it were not enough to be told by Copernicus that we are not in the center of the universe, by Charles Darwin that we have humble origins, and by Sigmund Freud that we are not full masters of our behavior, we have to concede that even in the realm of ethics there are forerunners and descent. But human ethical behavior has a degree of elaboration and complexity that makes it distinctly human. Ethical rules create uniquely human obligations for the normal individual acquainted with those rules. The codification is human; the narratives we have constructed around the situation are human. We can accommodate the realization that a part of our biological/ psychological makeup has nonhuman beginnings with the notion that our deep understanding of the human condition confers upon us a unique dignity.

Nor should the fact that our noblest cultural creations have forerunners imply that either humans or animals have a single and fixed social nature. There are varied kinds of social nature, good and bad, resulting from the vagaries of evolutionary variation, gender, and personal development. As Frans de Waal has

shown in his work, there are bad-natured apes, aggressive and ter-
ritorial chimpanzees, and good-natured apes, the bonobos, whose
wonderful personality resembles a marriage of Bill Clinton and
Mother Teresa.

The construction we call ethics in humans may have begun as
part of an overall program of bioregulation. The embryo of ethical
behaviors would have been another step in a progression that includes all the nonconscious, automated mechanisms that provide
metabolic regulation; drives and motivations; emotions of diverse
kinds; and feelings. Most importantly, the situations that evoke
these emotions and feelings call for solutions that include cooperation. It is not difficult to imagine the emergence of justice and
honor out of the practices of cooperation. Yet another layer of social emotions, expressed in the form of dominant or submissive
behaviors within the group, would have played an important role
in the active give and take that define cooperation.

It is reasonable to believe that humans equipped with this repertoire of emotions and whose personality traits include cooperative
strategies would be more likely to survive longer and leave more
descendants. That would have been the way to establish a genomic
basis for brains capable of producing cooperative behavior. This is
not to suggest that there is a gene for cooperative behavior, let alone
ethical behavior in general. All that would be necessary would be a
consistent presence of the many genes likely to endow brains with
certain regions of circuitry and with the attendant wiring—for example, regions such as the ventromedial frontal lobe that can interrelate certain categories of perceived events with certain emotion/
feeling responses. In other words, some genes working in concert
would promote the construction of certain brain components, and
the regular operation of those components, which, in turn, given
the appropriate environmental exposures, would make certain
kinds of cognitive strategy and behavior more probable under certain circumstances. In essence, evolution would have endowed our

brains with the apparatus necessary to recognize certain cognitive
configurations and trigger certain emotions related to the management of the problems or opportunities posed by those configurations. The fine tuning of that remarkable apparatus would depend
on the history and habitat of the developing organism.[19]

Lest it be thought that evolution and its baggage of genes has
simply made things wonderful by bringing to us all these proper
behaviors, let me point out that the nice emotions and the com-
mendable, adaptive altruism pertain to a *group*. In the animal
world, these groups include packs of wolves and troops of apes.
Among humans, they include the family, tribe, city, and nation.
For those outside the group, the evolutionary history of these responses shows that they have been less than kind. The nice emotions can easily turn nasty and brutish when they are aimed
outside the inner circles toward which they are naturally targeted.
The result is anger, resentment, and violence, all of which we can
easily recognize as a possible embryo of tribal hatreds, racism,
and war. This is the time to introduce the reminder that the best
of human behavior is not necessarily wired under the control of
the genome. The history of our civilization is, to some extent, the
history of a persuasive effort to extend the best of "moral senti-
ments" to wider and wider circles of humanity, beyond the re-
strictions of the inner groups, eventually encompassing the whole
of humanity. That we are far from finishing the job is easy to
grasp just by reading the headlines.

And there is some more natural darkness to contend with.
The trait of dominance—like its complement, submission—is an
important component of social emotions. Dominance has a positive face in that dominant creatures tend to provide solutions for
the problems of a community. They conduct negotiations and
lead the wars. They find the path to salvation along the roads that

lead to water, fruit, and shelter, or along the roads of prophecy and wisdom. But those dominant individuals also can become abusive bullies, tyrants, and despots, especially when dominance comes hand in hand with its evil twin: charisma. They can conduct negotiations wrongly and lead others to the wrong war. In those creatures, the display of nice emotions is reserved for an exceedingly small group made up of themselves and of those that sustain them most closely. Likewise, the submissive traits that can play such helpful roles in reaching agreement and consensus around a conflict also can make individuals cower under tyranny and hasten the downfall of an entire group by the sheer overuse of obeisance.

As conscious, intelligent, and creative creatures immersed in a cultural environment, we humans have been able to shape the rules of ethics, structure their codification into law, and design the application of the law. We will remain involved in that effort. The collective of interactive organisms, in a social environment and in the culture such a collective produces, are as important or more so in the understanding of these phenomena, even if the culture is itself conditioned to a large extent by evolution and neurobiology. To be sure, the beneficial role of the culture depends, in large measure, on the accuracy of the scientific picture of human beings the culture uses to forge its future path. And this is where modern neurobiology integrated in the traditional fabric of the social sciences may come to make a difference.

Largely for the same reasons, elucidating biological mechanisms underlying ethical behaviors does not mean that those mechanisms or their dysfunction are the guaranteed cause of a certain behavior. They may be determinative, but not *necessarily* determinative. The system is so complex and multilayered that it operates with some degree of freedom.

Not surprisingly, I believe that ethical behaviors depend on the workings of certain brain systems. But the systems are not centers—we do not have one or a few "moral centers." Not even the ventromedial prefrontal cortex should be conceived as a center. Moreover, the systems that support ethical behaviors are probably not dedicated to ethics exclusively. They are dedicated to biological regulation, memory, decision-making, and creativity. Ethical behaviors are the wonderful and most useful side effects of those other activities. But I see no moral center in the brain, and not even a moral system, as such.

On these hypotheses then, the grounding role of feelings is tied to their natural life-monitoring function. Ever since feelings began, their natural role would have been to keep the condition of life in mind and to make the condition of life count in the organization of behavior. And it is precisely because feelings continue to do so now that I also believe they should play a critical part in the current evaluation, development, and even application of the cultural instruments to which we have been alluding here.[20]

If feelings index the state of life within each living human organism, they also can index the state of life in any human group, large or small. Intelligent reflection on the relation between social phenomena and the experience of feelings of joy and sorrow seems indispensable for the perennial human activity of devising systems of justice and political organization. Perhaps even more importantly, feelings, especially sorrow and joy, can inspire the creation of conditions in the physical and cultural environments that promote the reduction of pain and the enhancement of well-being for society. In that direction, developments in biology and progress in medical technologies have bettered the human condition consistently over the past century. So have the sciences and technologies related to managing the physical environment. So have the arts, to some extent. So has the growth of wealth in democratic nations, to some extent.[21]

Homeostasis and the Governance of Social Life

Human life is first regulated by the natural and automatic devices of homeostasis—metabolic balance, appetites, emotions, and so forth. This most successful arrangement guarantees something quite astonishing: that *all* living creatures are given equal access to automatic solutions for managing life's basic problems, commensurate with their complexity and with the complexity of their niche in the environment. The regulation of our adult life, however, must go beyond those automated solutions because our environment is so physically and socially complex that conflict easily arises due to competition for resources necessary for survival and well-being. Simple processes such as obtaining food and finding a mate become complicated activities. They are joined by many other elaborate processes—think of manufacturing, commerce, and banking; health care, education, and insurance; and the numerous other support activities whose ensemble constitutes a human society with an economy. Our life must be regulated not only by our own desires and feelings but also by our *concern* for the desires and feelings of others expressed as social conventions and rules of ethical behavior. Those conventions and rules and the institutions that enforce them—religion, justice, and sociopolitical organizations—become mechanisms for exerting homeostasis at the level of the social group. In turn, activities such as science and technology assist the mechanisms of social homeostasis.

None of the institutions involved in the governance of social behavior tend to be regarded as a device to regulate life, perhaps because they often fail to do their job properly or because their immediate aims mask the connection to the life process. The ultimate goal of those institutions, however, is precisely the regulation of life in a particular environment. With only slight variations of accent, on the individual or the collective, directly or indirectly, the ultimate goal of these institutions revolves around promoting life and avoiding death and enhancing well-being and reducing suffering.

This was important for humans because automated life regulation can only go so far when the environments—not just physical but social—become exceedingly complex. Without the help of deliberation, pedagogy, or formal instruments of culture, nonhuman species exhibit useful behaviors that run from the trivial—finding food or a mate; to the sublime—showing compassion for another. But look, for a moment, at us humans. We certainly cannot dispense with any part of the gene-given innate apparatus of behavior. Yet it is apparent that, as human societies became more complex and certainly for the ten thousand or more years since agriculture was developed, human survival and well-being depended *on an additional kind of nonautomated governance* in a social and cultural space. I am referring to what we usually associate with reasoning and freedom of decision.[22] It is not just that we humans *show* compassion for another suffering being as bonobo chimpanzees and other nonhuman species can. We also *know that we feel* compassion, and, perhaps as a consequence, we have been doing something about the circumstances behind the events that provoked that emotion and feeling in the first place.

Nature has had millions of years to perfect the automated devices of homeostasis, while the nonautomated devices would have a history of a few thousand years. But I see other noticeable differences between automated and nonautomated life regulation. A major difference has to do with "goals" versus "ways and means." The goals and the ways and means of the automated devices are well established and efficacious. However, when we turn to the nonautomated devices, we see that while some goals have been largely agreed upon—such as not killing the other—many goals are still open to negotiation and remain to be established—how exactly to help the sick and needy. Moreover, the ways and means to reach any goals have varied remarkably with the human group and the historical period, and are anything but fixed. Feelings may have contributed to articulating the goals that define humanity at its most refined—not harming others; promoting the good of another. But the story of humanity is one of struggle to find acceptable ways and means to implement those goals. One might say that the goals of Marxism, albeit narrow, were laudable in some respects since the stated intention was to create some kind of fair world. Yet the ways and means of the societies that promoted Marxism were disastrous because, among other reasons, they were in frequent clash with well-established mechanisms of *automated* life regulation. The good of the larger collective often required the pain and suffering of many individuals. The result was a costly human tragedy. The incipience and fragility of the nonautomated devices is easily demonstrated by Nazism, in which both goals and ways and means were deeply flawed. In most respects, then, the nonautomated devices are a work in progress, still hampered by the enormous difficulty of negotiating goals and finding ways and means that do not violate other aspects of life regulation. From this perspective, I believe feelings remain essential to maintaining those goals the cultural group considers inviolable and worthy of perfecting. Feelings also are a necessary guide to the invention and negotiation of ways and means that, somehow, will not clash with basic life regulation and distort the intention behind the goal. Feelings remain as important today as when humans first discovered that killing other humans was a questionable action.

Social conventions and ethical rules may be seen in part as extensions of the basic homeostatic arrangements at the level of society and culture. The outcome of applying the rules is the same as the outcome of basic homeostatic devices such as metabolic regulation or appetites: a balance of life to ensure survival and well-being. The extension does not stop there, however. It reaches into the larger organizational levels of which social groups are a part. The constitution that governs a democratic state, the laws that are consonant with that constitution, and application of those laws in a judicial system also are homeostatic devices. They are linked by a long umbilical cord to the other tiers of homeostatic regulation on which they are modeled: appetites/desires, emotions/feelings, and the conscious governance of both. So is the fledgling, twentieth-century development of worldwide bodies of social coordination, such as the World Health Organization, UNESCO, and the much-maligned United Nations. All of these institutions can be seen as part and parcel of the tendency to promote homeostasis on a large scale. Along with the good results they often achieve, however, these bodies suffer from many ills and their policies are often informed by deficient conceptions of humanity that have not taken into account emerging scientific evidence. Still, their imperfect presence is a sign of progress and a beacon of hope, however weak. And there are other reasons for hope. The study of social emotions is in its infancy. If the cognitive and neurobiological investigations of emotions and feelings can join forces with, for example, anthropology and evolutionary psychology, it is likely that some of the suggestions contained in this chapter can be tested. We might get a glimpse of how human biology and culture really mesh behind the appearance, and we might even be able to guess how the genome and the physical and social environments interacted during the long history of evolution.

The foregoing, I note again, are ideas whose merits remain to be assessed. A formal proposal on the neurobiology of ethical behaviors is outside the scope of this book, and so is a discussion of these ideas in a historical perspective.[23]

The Foundation of Virtue

I wrote early in this book that my return to Spinoza came almost by chance as I tried to check the accuracy of a quote I kept on a yellowed paper, a link to the Spinoza I had read long ago. Why had I kept the quote? Perhaps because it was something I intuited as specific and illuminating. But I had never paused to analyze it in detail until it traveled from my memory to the page I was working on.

The quote comes from Proposition 18 in part IV of *The Ethics* and it reads: "... the very first foundation of virtue is the endeavor (conatum) to preserve the individual self, and happiness consists in the human capacity to preserve its self." In Latin the proposition reads *... virtutis fundamentum esse ip sum conatum proprium esse conservandi, et felicitatem in eo consistere, quòd homo suum esse conservare potest.* A comment on the terms used by Spinoza is in order before we go any further. First, as noted earlier, the word *conatum* can be rendered as endeavor or tendency or effort, and Spinoza may have meant any of these, or perhaps a blend of the three meanings. Second, the word *virtutis* can refer not just to its traditional moral meaning, but also to power, and ability to act. I shall return to this issue. Curiously, in this passage, he uses the word *felicitatem,* which is best translated as happiness, rather than *laetitia,* which can be translated as joy, elation, delight, *and* happiness.

At first glance the words sound like a prescription for the selfish culture of our times but nothing could be further from their real meaning. As I interpret it, the proposition is a cornerstone for a generous ethical system. It is an affirmation that at the base of whatever rules of behavior we may ask humanity to follow, there is something inalienable: A living organism, known to its owner because the owner's mind has constructed a self, has a natural tendency to preserve its own life; and that same organism's state of optimal functioning, subsumed by the concept of joy, results from the successful endeavor to endure and prevail. Paraphrased in deeply American terms I would rewrite Spinoza's proposition as follows: I hold these truths to be self-evident, that all humans are created such that they tend to preserve their life and seek well-being, that their happiness comes from the successful endeavor to do so, and that the foundation of virtue rests on these facts. Perhaps these resonances are not a coincidence.

Spinoza's statement rings clear as a bell, but it does require elaboration for its full impact to be appreciated. Why should a concern for oneself be the basis for virtue, lest that virtue pertain to that self alone? Or, to put it more bluntly, how does Spinoza move from oneself to all the selves to whom virtue must apply? Spinoza makes the transition relying again on biological facts. Here is the procedure: The biological reality of self-preservation leads to virtue because in our inalienable need to maintain ourselves we must, of necessity, help preserve *other* selves. If we fail to do so we perish and are thus violating the foundational principle, and relinquishing the virtue that lies in self-preservation. The secondary foundation of virtue then is the reality of a social structure and the presence of other living organisms in a complex system of interdependence with our own organism. We are in a bind, literally, in the good sense of the word. The essence of this transition can be found in Aristotle, but Spinoza ties it to a biological principle—the mandate for self-preservation.

So here is the beauty behind the cherished quote, seen from today's perspective: It contains the foundation for a system of ethical behaviors and that foundation is neurobiological. The foundation is the result of a discovery based on the observation of human nature rather than the revelation of a prophet.

Human beings are as they are—living and equipped with appetites, emotions, and other self-preservation devices, including the capacity to know and to reason. Consciousness, in spite of its limitations, opens the way for knowledge and reason, which, in turn, allow individuals to discover what is good and evil. Again, good and evil are not revealed, they are discovered, individually or by agreement among social beings.

The definition of good and evil is simple and sound. Good objects are those that prompt, in reliable and sustainable fashion, the states of joy that Spinoza sees as enhancing the power and freedom of action. Evil objects are those that elicit the opposite result: Their encounters with an organism are disagreeable to that organism.

And what about good and evil actions? Good actions and evil actions are not merely actions that do or do not accord with individual appetites and emotions. Good actions are those that, while producing good for the individual via the natural appetites and emotions, *do not harm other individuals*. The injunction is unequivocal. An action that might be personally beneficial but would harm others is not good because harming others always haunts and eventually harms the individual who causes the harm. Consequently such actions are evil. "... our good is especially in the friendship that links to other humans and to advantages for society" (*The Ethics*, Part IV, Proposition 10). I interpret Spinoza to mean that the system constructs ethical imperatives based on the presence of mechanisms of self-preservation in each person, but mindful of social and cultural elements as well. Beyond each self there are *others*, as individuals and as social entities, and their own self-preservation, e.g., their appetites and emotions, must be taken into consideration. Neither the essence of the *conatus*, nor the notion that harm to the other is harm to the self are Spinoza's inventions. But perhaps the Spinozian novelty resides with the powerful blend of the two.

The endeavor to live in a shared, peaceful agreement with others is an extension of the endeavor to preserve oneself. Social and political contracts are extensions of the personal biological mandate. We happen to be biologically structured in a certain way—mandated to survive and to maximize pleasurable rather than painful survival—and from that necessity comes a certain social agreement. It is reasonable to hypothesize that the tendency to seek social agreement has itself been incorporated in biological mandates, at least in part, due to the evolutionary success of populations whose brains expressed cooperative behaviors to a high degree.

Beyond basic biology there is a human decree which is also biologically rooted but arises only in the social and cultural setting, an intellectual product of knowledge and reason. Spinoza sensed this arrangement clearly: "For example, the law that all bodies impinging on lesser bodies, lose as much of their own motion as they communicate to the latter is a universal law of all bodies, and depends on natural necessity. So, too, the law that a man in remembering one thing straightway remembers another either like it, or which he had perceived simultaneously with it, is a law that necessarily follows from the nature of man. But the law that men must yield, or be compelled to yield, somewhat of their natural right, and that they bind themselves to live in a certain way, depends on human decree. Now, though I freely admit that all things are predetermined by universal natural laws to exist and operate in a given, fixed, and definite manner, I still assert that the laws I have just mentioned depend on human decree."[24]

Spinoza would have been pleased to know that one reason why the human decree may take cultural roots is that the design of the human brain tends to facilitate its practice. It is likely that the simplest form of some behaviors necessary for realizing the human decree, such as reciprocal altruism and censure, is merely waiting to be awakened by social experience. We have to work hard at formulating and perfecting the human decree but to some extent our brains are wired to cooperate with others in the process of making the decree possible. This is the good news. The bad news, of course, is that many negative social emotions, along with their exploitation in modern cultures, make the human decree difficult to implement and improve.

The importance of the biological facts in the Spinoza system cannot be overemphasized. Seen through the light of modern biology, the system is conditioned by the presence of life; the presence of a natural tendency to preserve that life; the fact that the preservation of life depends on the equilibrium of life functions and consequently on life regulation; the fact that the status of life regulation is expressed in the form of affects—joy, sorrow—and is modulated by appetites; and the fact that appetites, emotions, and the precariousness of the life condition can be known and appreciated by the human individual due to the construction of self, consciousness, and knowledge-based reason. Conscious humans know of appetites and emotions as feelings, and those feelings deepen their knowledge of the fragility of life and turn it into a *concern*. And for all the reasons outlined above the concern overflows from the self to the other.

I am not suggesting that Spinoza ever said that ethics, law, and political organization were homeostatic devices. But the idea is compatible with his system considering the way he saw ethics, the structure of the state, and the law as means for individuals to achieve the natural balance expressed in joy.

It is often said that Spinoza did not believe in free will, a notion that appears to be in direct conflict with an ethical system in which human beings decide to behave in a particular way, according to clear imperatives. But Spinoza never denied that we are aware of making choices and that, for all intents and purposes, we *can* make choices, and willfully control our behavior. He constantly recommended that we forego whatever action we consider wrong in favor of one we consider right. His entire strategy for human salvation depends on our making deliberate choices. The issue for Spinoza is that many seemingly deliberate behaviors can be explained by prior conditions of our biological constitution, and that, ultimately, everything we think and do results from certain antecedent conditions and processes that we may not be able to control. But we still can say a categorical no, just as firmly and imperatively as Immanuel Kant would, however illusory the freedom of that no may be.

There is an additional meaning to Spinoza's Proposition 18. It hinges on the double meaning of the word virtue, on the emphasis given to the notion of happiness, and on the many comments that follow in Parts IV and V of *The Ethics*. Some degree of happiness comes quite simply from acting in conformity with our self-preserving tendency, as needed but not more. In addition to urging the establishment of a social contract, Spinoza is telling us that happiness is the power to be free of the tyranny of negative emotions. Happiness is not a reward for virtue: it is virtue itself.

What Are Feelings For?

So, why do we have feelings? What do feelings accomplish for us? Would we be better off without them? These questions have been perennially deemed unanswerable, but I believe we can begin to address them now. For one thing, we have a workable idea of what feelings are and that is a first step in the attempt to discover *why* feelings are and *what* they do for us. For another, we have just seen how the partnership of emotion and feeling plays a critical role in social behavior and, by extension, in ethical behavior. Still, a skeptic might remain unconvinced and argue that nonconscious emotion alone would suffice to guide social behavior; or that neural mappings of emotional states would suffice, without any need for those maps to become mental events, i.e., feelings.

In short, there would be no need for mind, let alone a conscious mind. Let me attempt to answer the skeptics.

The answer to "why" begins as follows. In order for the brain to coordinate the myriad body functions on which life depends, it needs to have maps in which the state of varied body systems are represented moment by moment. The success of this operation depends on this massive mapping. It is critical to know what is going on in different body sectors so that certain functions can be slowed down, halted, or called into action, and so that appropriate corrections in the governance of the organism's life can be made. Examples of the situation I have in mind include a local wound, externally inflicted or caused by infection; or the malfunction of an organ such as the heart or kidneys, or a hormonal imbalance.

Neural maps that are critical for the governance of life turn out to be a necessary basis for the mental states we call feelings. This takes us one more step into the answer to the "why" question: feelings probably arose as a by-product of the brain's involvement in the management of life. Had there been no neural maps of body state there might never have been such a thing as feelings.

These answers may raise some objections. For example, it can be argued that since the basic processes of life governance are automated and nonconscious, feelings, which in the usual sense of the term are conscious, would be superfluous. A skeptic would say that the brain can coordinate life processes and execute physiological corrections on the basis of the neural maps alone, without any help from conscious feelings. The mind would not need to *know* about the contents of those maps. This argument is only correct in part. True enough, to a certain extent, body-state maps assist the brain with the management of life even when the "owner" of the organism does not know such maps exist. But the objection misses an important point made earlier. Body-state

maps can provide only limited assistance without conscious feelings. The maps work for problems of a certain degree of complexity and no more; when the problem gets too complicated—
when it requires a mixture of automated responses *and* reasoning
on accumulated knowledge—unconscious maps no longer help
and feelings come in handy.

What does the feeling level add to problem-solving and
decision-making that the neural-map level of those events, as currently described by neuroscience, cannot offer? In my view there
are two tiers to the answer, one that has to do with the status of
feelings as mental events in the conscious mind, another that has
to do with what feelings stand for.

The fact that feelings are mental events is relevant for the following reason. Feelings help us solve nonstandard problems involving creativity, judgment, and decision-making that require
the display and manipulation of vast amounts of knowledge. Only
the "mental level" of biological operations permits the timely integration of large sets of information necessary for the problem-
solving processes. Because feelings have the requisite mental
level, they can enter the mind fray and influence the operations.
At the end of chapter 5, I will return to the issue of what the mental level of neural processing brings to organisms that other levels
do not.

What feelings bring to the mind fray is just as important.
Conscious feelings are prominent mental events that call attention to the emotions that begat them, and to the objects that
triggered those emotions. In individuals who also have an autobiographical self—the sense of personal past and anticipated future also known as extended consciousness—the state of feeling
prompts the brain to process emotion-related objects and situations *saliently*. The appraisal process that led to the isolation of the
object and the onset of the emotion can be revisited and analyzed

as needed. Moreover, conscious feelings also call attention to the consequences of the situation: What is the object that triggered the emotion about to do? How did the causative object affect the person who feels? What are the person's thoughts now? Occurring in an autobiographical setting, feelings generate a concern for the individual experiencing them. The past, the now, and the anticipated future are given the appropriate saliencies and a better chance to influence the reasoning and decision-making process.

When feelings become known to the self within the organism that possesses them, feelings improve and amplify the process of managing life. The machinery behind feelings enables the biological corrections necessary for survival by offering explicit and highlighted information as to the state of different components of the organism at each given moment. Feelings label the related neural maps with a stamp that reads: "Mark that!"

One might summarize by saying that feelings are necessary because they are a mental-level expression of emotions and what lies beneath them. Only at that mental level of biological processing and in the full light of consciousness is there sufficient integration of the now, the past, and the anticipated future. Only at that level is it possible for emotions to create, via feelings, the concern for the individual self. The effective solution of nonstandard problems requires the flexibility and high power information gathering that mental processes can offer, as well as the mental concern that feelings can provide.

The process of learning and recalling emotionally competent events is different with conscious feelings from what it would be without feelings. Some feelings optimize learning and recall. Other feelings, extremely painful feelings in particular, perturb learning and protectively suppress recall. In general, memory of the felt situation promotes, consciously or not, the avoidance of events associated with negative feelings *and* the seeking of situations that may cause positive feelings.[25]

We should not be surprised that the neural machinery underlying feelings prevailed sturdily in evolution. Feelings are not superfluous. All that gossip from deep within turns out to be quite
useful. It is not a simple issue of trusting feelings as the necessary
arbiter of good and evil. It is a matter of discovering the circumstances in which feelings can indeed be an arbiter, and using the
reasoned coupling of *circumstances and feelings* as a guide to human behavior.

CHAPTER 5

Body, Brain, and Mind

Body and Mind

Are mind and body two different things or just one? If they are not the same, are mind and body made from two different substances or just one? If there are two substances, does the mind substance come first and cause the body and its brain to exist, or does the body substance come first and its brain cause the mind? Also, how do those substances interact? Now that we understand in some detail how neural circuits operate, how is it that the activity of those circuits relates to the mental processes we can introspect? These are some of the main issues involved in the so-called mind-body problem, a problem whose solution is central to the understanding of who we are. In the eyes of many scientists and philosophers, the problem is either false or has been resolved. But in relation to the questions posed above, general agreement is confined to the notion that the mind is a process, not a thing. When perfectly reasonable, intelligent, and educated people can disagree passionately about these issues, the least that can be said is that the solution either is not satisfactory or has not been satisfactorily presented.

Until recently the mind-body problem remained a philosophical topic, outside the realm of empirical science. Even in the twentieth century, when it appeared that the time had come for the sciences of mind and brain to deal with the problem, the barriers raised were so many—in terms of method and approach— that the issue was postponed once again. Only during the past decade has the problem finally entered the scientific agenda, largely as a part of the investigation of consciousness. It is important to note, however, that *consciousness* and *mind* are not synonymous. In the strict sense, consciousness is the process whereby a mind is imbued with a reference we call self, and is said to know of its own existence and of the existence of objects around it. Elsewhere I have explained that in certain neurological conditions there is evidence that the mind process continues, but consciousness is impaired. *Consciousness* and *conscious mind,* however, are synonymous.[1]

Neurobiological and cognitive studies have elucidated some aspects of the mind-body puzzle, but the resulting interpretations remain so contested that there is little incentive for reflection on the existing evidence or the gathering of new evidence. This is unfortunate because, in spite of the barriers, progress is being made, and there is more knowledge available than meets the eye, if only the eye is theoretically free to see.[2]

At this point in the book it is appropriate to consider the mind-body problem for two reasons. First, much of what I have proposed regarding emotion and feeling is especially pertinent to the debate about the mind-body problem. Second, the problem is central to Spinoza's thinking. In fact, Spinoza may have gleaned part of a solution, a possibility that, correctly or incorrectly, added strength to my own convictions on this issue. Perhaps for that very reason I remember the place and time around which I first consolidated my current views of the problem. The place was The Hague and the occasion a Huygens Lecture I was invited to give.

The Hague, December 2, 1999

The annual Huygens Lecture is named after Christiaan Huygens. Huygens had little to do with brain or mind or philosophy and everything to do with astronomy and physics. He cared for space:

He discovered Saturn's rings and estimated the distance between the earth and the stars by pin-holing the sun. He cared for time: He invented the pendulum clock. And he cared for light: The Huygens principle refers to his wave theory of light. As the most famous scientist in Holland's history, he is the patron saint of this annual lecture, which is meant to feature any field of science. By the way, in his time Huygen's father, Constantijn, was as famous as the son and no less remarkable. His knowledge spanned Latin, music, mathematics, literature, history, and the law. He was a well-rounded art connoisseur. He was a poet. He was a statesman—a secretary to Holland's stadtholder, as his own father had been. The pressing task of filling the palaces of the state with suitable paintings also made him a patron of the arts. His great discovery: Rembrandt.

The subject of my lecture is the neural basis of the conscious mind, and given the drift of my thoughts over the past year, the connection to Huygens turns out to be quite appropriate. Huygens and Spinoza were contemporaries. They were born not quite three years apart and were even neighbors for a time. Of course, Huygens lived in splendor, not in a rented apartment—the Huygens family had a palace in The Hague and a large estate between

The Hague and Voorburg. But they did breathe the same air and saw each other on several occasions. Huygens acquired lenses from Spinoza and wrote to him from time to time with questions about philosophical issues. Spinoza was well acquainted with Huygens's work and owned copies of his books. At least three letters from Spinoza to Huygens survive, dated 1666, replying to Huygens's questions on the unity of God. They are addressed "Distinguished Sir," and behind the matter-of-fact tone one senses more than an arm's length. Spinoza goes straight to the point, wasting no time with ceremonial niceties. The worlds of the now outcast Dutch-Jew and the established Dutch aristocrat might have been bridged by their intellectual curiosity, but their personalities seem to have been too different for any friendship to be possible. Nonetheless, they knew where each other stood. Huygens knew that Spinoza had little patience for Huygens's sometime teacher, René Descartes—Descartes had initiated young Christiaan in the mysteries of algebra—and that was well and good because Huygens had grown almost as disenchanted with Descartes' ideas as Spinoza, though not for quite the same reasons. Huygens may have referred to Spinoza as the "Jew of Voorburg" or "our Israelite," but he thought the lenses Spinoza produced were the best and he respected Spinoza's intellect enough to see him as potential competition. From Paris, where he lived for extensive periods and where he comfortably sat out most wars involving the Dutch, Huygens would write home to his brother advising him not to share new ideas with Spinoza. The coolness was mutual.

The Huygens Lecture is given in the New Church, the distinctive seventeenth-century landmark, a few yards from Spinoza's grave and a few blocks from Spinoza's home.[3] As I speak I am distracted by the thought of Spinoza, somewhat buried behind me, to the left, somewhat living behind me, to the right. I dutifully deliver the lecture I planned, but my thoughts are caught up with the idea that Spinoza may have prefigured some of the conclusions I am about to present.

The Invisible Body

It is easy to understand why the mind may appear to be a forbidding, unapproachable mystery. The mind, as an entity, seems to be different in kind from other things we know, namely, from the objects around us and from the parts of our own bodies that we see and touch. The view of the mind-body problem known as substance dualism captures that first impression: The body and its parts are physical matter while the mind is not. When we let a part of our minds observe the rest of our minds, naturally and naïvely, without the influence of the scientific knowledge currently available, the observations appear to reveal, on the one hand, the physically extensive matter that constitutes the cells, tissues, and organs of our bodies. On the other, they reveal the stuff that we cannot touch—all the rapidly formed feelings, sights, and sounds that make up the thoughts in our minds, which we assume, without any evidence for or against, to be another kind of substance, a nonphysical one.

The view of the mind-body problem that results from these uninformed reflections splits the mind to one side and the body and its brain to the other. This view, substance dualism, is no longer mainstream in science or philosophy, although it is probably the view that most human beings today would regard as their own.

In its outline, the substance dualism view is the account that Descartes helped dignify and which is difficult to reconcile with his remarkable scientific achievements. Descartes was ahead of his peers in the way he conceived of complicated mechanisms for the operations of the body. He broke with the scholastic tradition by weaving together two worlds that remained separate until then: the physical-inorganic and the living-organic. He was equally adroit at conceiving sophisticated operations for the mind and insisted that mind and body influenced each other mutually. Yet he never proposed a plausible means for those mutual influences to exert themselves. In a bizarre twist, Descartes proposed that mind and body interacted, but never explained how the interaction might take place beyond saying that the pineal gland was the conduit for such interactions. The pineal is a small structure, located at the midline and base of the brain, and it turns out to be rather poorly connected and endowed for the momentous job Descartes required of it. In spite of Descartes' sophisticated views of mental and physiological body processes, which he separately considered, he either left the mutual connections of mind and body unspecified or made them implausible. Princess Elizabeth of Bohemia, the sort of bright and friendly student we all wish to have, saw quite clearly then what we see clearly now: For mind and body to do the job Descartes required of them, mind and body needed to make *contact*. However, by emptying mind of any physical property, Descartes made contact impossible.[4]

For Descartes, the human mind remained devoid of spatial extension and material substance, two negative traits that made it capable of living after the body no longer existed. It was a substance but it was not physical. Whether Descartes really believed this formulation is by no means certain. He might have believed it, at some point, and then not, which is not meant at all as a criticism. It would simply mean that Descartes was uncertain and ambivalent about a conception that has chronically plunged human beings, learned as well as ignorant, smart as well as stupid, into precisely the same state of uncertainty and ambivalence. Very human and very understandable. Believed or not, however, the fact that his formulation reaffirmed the immortality of the personal mind allowed him to escape the anathema that was to befall Spinoza only a few years later. Unlike Spinoza, Descartes has been acknowledged continuously well into our own era, by philosophers, scientists, and the lay public, although not always favorably.

In spite of its scientific shortcomings, the view identified with Descartes resonates well with the awe and wonder we deservedly have for our own minds. There is no doubt that the human mind is special—special in its immense capacity to feel pleasure and pain and to be aware of the pain and pleasure of others; in its ability to love and to pardon; in its prodigious memory; in its ability to symbolize and narrate; in its gift of language with syntax; in its power to understand the universe and create new universes; in the speed and ease with which it processes and integrates disparate information so that problems can be solved. But awe and wonder at the human mind are compatible with other views of the relation between body and mind and do not make Descartes' views any more correct.

As the observations made possible by introspection became increasingly informed by the modern scientific facts of neurobiology, the substance dualistic view of the mind-body problem lost its appeal. Mental phenomena were revealed as closely dependent on the operation of many specific systems of brain circuits. For example: Seeing depends on several specific neural regions located along pathways from the retina to the cerebral hemispheres. When one of those regions is removed, vision is disturbed. When *all* of the vision-related neural regions are removed, vision is compromised in its entirety. Likewise for hearing, smelling, moving, speaking, or whatever high mental function you fancy. Even minor perturbations of the specific neural systems entail a major modification of mental phenomena. Perturbations caused by circumscribed damage to the nerve cells of certain neural regions—as occurs in a stroke, which causes a lesion—markedly change the content and form of feelings and thoughts. As we have seen, the same happens following temporary chemical and pharmacological alterations in the function of those nerve cells due to the administration of a drug, even when no permanent damage occurs. And so, perhaps for most scientists working on mind and brain, the fact that the mind depends closely on the workings of the brain is no longer in question. We can all celebrate the prescience of Hippocrates, who held the very same view, all by himself, a couple of millennia ago.

Uncovering a causative nexus from brain to mind, and a dependence of mind on brain, is good news, of course, but we should recognize that we have not yet elucidated the mind-body problem satisfactorily, and that the enterprise faces several hurdles, large and small. At least one of those hurdles could be overcome with a simple change of perspective. The hurdle relates to a curious situation: While the modern scientific coupling of brain and mind is most welcome, it does not do away with the dualistic split between mind and body. It simply shifts the position of the split. In the most popular and current of the modern views, the mind and brain go together, on one side, and the body (that is, the entire organism minus the brain) goes on the other side. Now the split separates brain and "body-proper" and the explanation of how mind and brain are related becomes more difficult when the brain-part of the body is divorced from the body-proper. Regrettably, this dualistic frame still works like a screen and does not let us see what is clearly in front of our eyes—namely the body in its broadest sense, and its relevance to the making of the mind.

This invisible body reminds me of Chesterton's invisible man.[5] You may know the story. A much foretold murder was committed inside a house while four people stood guard and closely watched who was coming and going from the house. That this fully expected murder came to pass was not a puzzle. The puzzle was that the victim was alone and the four observers were adamant: No one had gone in or out of the house. But this was quite false: The postman had gone into the house, done the deed, and left the house in plain view. He had even left unhurried footprints in the snow. Of course, everyone had looked at the postman, and yet all claimed not to have seen him. He simply did not fit the theory they had formulated for the identity of the possible murderer. They were looking but not seeing.

I am afraid something comparable has been happening with the great mystery behind the mind-body problem. Working toward a solution, even a partial solution, requires a change in perspective. It requires an understanding that the mind arises from or in a brain situated within a body-proper with which it interacts; that due to the mediation of the brain, the mind is grounded in the body-proper; that the mind has prevailed in evolution because it helps maintain the body-proper; and that the mind arises from or in biological tissue—nerve cells—that share the same characteristics that define other living tissues in the body-proper. Changing the perspective will not in itself solve the problem, but I doubt that we can get to the solution without changing the perspective.

Losing the Body and Losing the Mind

Sometimes one is struck by observations that change one's way of thinking. Sometimes the opposite occurs and one is struck by how one's current thinking changes the significance of a prior observation. Sometimes, if one is lucky, the reassessment of an observation actually focuses one's thinking. The latter happened to me regarding a certain patient I saw as a young neurologist. Pointing to his own body with precision, the patient described a strange sensation that would begin in the pit of his stomach and then rise into his chest, whereupon he would lose the feeling of the body below that level as if he were under local anesthesia. The anesthesia-like feeling would continue to rise and by the time it would reach up to his throat, he would pass out.

The patient was describing the upward march of a distortion in the sensation of his body, immediately followed by a complete loss of consciousness when the feeling of his body had gone from strange to entirely absent. A few moments after these momentous events, unbeknownst to him, he would be shaken by convulsions, as part of his epileptic fit. Minutes later, the seizure over, the patient would be returned to his normal life.

It is common for patients with epilepsy to describe strange sensations prior to the onset of the seizures. These phenomena are called *auras,* and auras such as this patient's, which begin near the stomach or the lower chest, are called "epigastric." They are one of the most common varieties of the phenomenon. Patients often report the rise of these strange sensations from the abdomen to the neck, followed by the loss of consciousness.[6]

Why did this patient's unexceptional story become significant for me? It was because, long after it happened, the case raised this possibility: When the ongoing brain mapping of the body was suspended so was the mind. In a way, removing the mental presence of the body was like pulling the rug from under the mind. A radical interruption in the flow of body representations that support our feelings and our sense of continuity might entail, in and of itself, a radical interruption of our thoughts of objects and situations.[7]

Many years later, seeing a patient with a condition known as asomatognosia, the suggestion outlined above became more plausible. In that patient most but not all of the body sensation would gradually vanish over a very brief period and remain so for several minutes, yet mind and self were *not* suspended. The sense of the body frame and its musculature disappeared, in both the trunk and the limbs, but the sense of the viscera, namely that of the heart beating, remained. The patient stayed awake and alert during the unfolding of these disquieting episodes, although she could not will herself to move and could think of nothing else but her unusual condition. It was hardly a normal state of mind, obviously, but still mind enough to observe and report on that commotion. In the patient's vivid description, "I didn't lose any sense of being, just lost my body," although, to be precise, she should have said that she lost part of her body. The condition raised the possibility that as long as there is *some* body representation—as long as the rug was not pulled completely from under the mind— the mind process could be grounded. It also raised the possibility that some body representations may be of greater value than others to ground the mind, namely, those that pertain to the organism's interior, specifically to the viscera and internal milieu. Incidentally, the patient's condition was caused by a previous stroke that had compromised one of the somatosensory regions of her right cerebral hemisphere and created a small area of scarred brain tissue. This tissue was the source of a local epileptic seizure, an advancing electrical wave that disturbed the function of some body-mapping circuits temporarily. We suspect that maps in SII, SI, and perhaps the right angular gyrus were rendered dysfunctional during the seizure, but that the insula was spared.

Through the years I have remained intrigued by the rare situations in which the perception of parts of the body is modified by disease. If only one limb is involved things can get strange enough. For example, a limb whose nerves have been sectioned may feel distorted, out of place, or absent; and a limb that has been amputated may feel quite present, thanks to a phantom. Not a good state of affairs, but tolerable in the long run.[8] When the perception of *extensive* sectors of the body is disturbed, however, even if temporarily, the cost to the patient is always some measure of mental disarray. The underlying mechanism always involves one of the body-sensing regions or body-related pathways discussed in Chapter Three. Cases involving the body-signaling pathways are the rarest of all because there are so many signaling routes from body to brain that neurological diseases are unlikely to compromise most of them.[9]

I cannot say that my current perspective on the mind-body problem was based on the above facts. Still these facts, along with the findings on emotion and feeling discussed in chapters 2 and 3, focused my thinking and helped me reconcile a theoretical account with a human reality. In brief, the theoretical account specifies the following:

- That the body (the body-proper) and the brain form an integrated organism and interact fully and mutually via chemical and neural pathways.
- That brain activity is aimed primarily at assisting with the regulation of the organism's life processes both by coordinating internal body-proper operations, and by coordinating the interactions between the organism as a whole and the physical and social aspects of the environment.
- That brain activity is aimed primarily at survival with well-being; a brain equipped for such a primary aim can engage in anything else secondarily from writing poetry to designing spaceships.
- That in complex organisms such as ours, the brain's regulatory operations depend on the creation and manipulation of mental images (ideas or thoughts) in the process we call mind.
- That the ability to perceive objects and events, external to the organism or internal to it, requires images. Examples of images related to the exterior include visual, auditory,

tactile, olfactory, and gustatory images. Pain and nausea
are examples of images of the interior. The execution of
both automatic and deliberated responses requires images. The anticipation and planning of future responses
also require images.

- That the critical interface between body-proper activities
and the mental patterns we call images consists of specific brain regions employing circuits of neurons to construct continual, dynamic neural patterns corresponding
to different activities in the body—in effect, mapping
those activities as they occur.

- That the mapping is not necessarily a passive process.
The structures in which the maps are formed have their
own say on the mapping and are influenced by other
brain structures.

Because the mind arises in a brain that is integral to the organism, the mind is part of that well-woven apparatus. In other
words, body, brain, and mind are manifestations of a single organism. Although we can dissect them under the microscope, for
scientific purposes they are in effect inseparable under normal
operating circumstances.

The Assembly of Body Images

From my perspective the brain produces two kinds of images of
the body. The first I call *images from the flesh*. It comprises images
of the body's interior, drawn, for example, from the sketchy neural
patterns that map the structure and state of viscera such as the
heart, the gut, and the muscles, along with the state of numerous
chemical parameters in the organism's interior.

The second kind of body image concerns particular parts of
the body, such as the retina in the back of the eye and the cochlea

in the inner ear. I call these *images from special sensory probes*. They are images based on the state of activity in those particular body parts when they are modified by objects that physically impinge upon those devices from outside the body. That physical impingement takes many forms. In the case of the retina and cochlea, respectively, the objects perturb the patterns of light and sound waves, and the altered pattern is captured in the sensory devices. In the case of touch, the actual mechanical contact of an object against the body boundary will change the activity of nerve endings distributed in the boundary itself—the skin. Shape and texture images are derivatives of this process.

The range of body changes that can be mapped in the brain is very wide. It includes the microscopic changes that occur at the level of chemical and electrical phenomena (for example, in the specialized cells of the retina that respond to patterns of photons carried in light rays). It also includes macroscopic changes that can be seen by the naked eye (a limb moving) or sensed at the tip of a finger (a bump in the skin).

In either body image, from the flesh or from the special sensory probes, the mechanism of production is the same. First, the activity in body structures results in momentary structural body changes. Second, the brain constructs maps of those body changes in a number of appropriate regions with the help of chemical signals conveyed in the bloodstream, and electrochemical signals conveyed in nerve pathways. Finally, the neural maps become mental images.

In the first kind of body images, the images from the flesh, the changes occur all over our interior landscape and are signaled to the body-sensing regions of the central nervous system by chemical molecules and nerve activity. In the second kind of body image, the images from special sensory probes, the changes occur within highly specialized body parts such as the retina. The resulting signals are relayed by neuronal connections to regions dedicated to mapping the state of that specialized body receptor. The regions are made of collections of neurons whose state of activity or inactivity forms a pattern that can be conceived as a map or representation of whatever event caused the activity to occur at a given time in a certain group of neurons and not in another. In the case of the retina, for example, those vision-related structures include the geniculate nucleus (part of the thalamus), the superior colliculus (part of the brain stem), and the visual cortices (part of the cerebral hemispheres). The list of specialized parts of the body includes: the cochlea in the inner ear (related to sound); the semicircular canals of the vestibule, also within the inner ear, where the vestibular nerve begins (the vestibule is related to the mapping of the body's position in space; our sense of balance depends on it); the olfactory nerve endings in the nasal mucosae (for the sense of smell); the gustatory papillae in the back of the tongue (for taste); and the nerve endings distributed in the superficial layers of the skin (for touch).

I believe that the foundational images in the stream of mind are images of some kind of body event, whether the event happens in the depth of the body or in some specialized sensory device near its periphery. The basis for those foundational images is a collection of brain maps, that is, a collection of patterns of neuron activity and inactivity (neural patterns, for short) in a variety of sensory regions. Those brain maps represent, comprehensively, the structure and state of the body at any given time. Some maps relate to the world within, the organism's interior. Other maps relate to the world outside, the physical world of objects that interact with the organism at specific regions of its shell. In either case, what ends up being mapped in the sensory regions of the brain and what emerges in the mind, in the form of an idea, corresponds to some structure of the body, in a particular state and set of circumstances.[10]

A Qualification

It is important that I qualify these statements, especially the last. There is a major gap in our current understanding of how neural patterns become mental images. The presence in the brain of dynamic neural patterns (or maps) related to an object or event is a *necessary* but not sufficient basis to explain the mental images of the said object or event. We can describe neural patterns—with the tools of neuroanatomy, neurophysiology, and neurochemistry—and we can describe images with the tools of introspection. How we get from the former to the latter is known only in part, although the current ignorance neither contradicts the assumption that images are biological processes nor denies their physicality. Many recent studies on the neurobiology of consciousness address this issue. Most consciousness studies are actually centered on this issue of the making of the mind, the part of the consciousness puzzle that consists of having the brain make images that are synchronized and edited into what I have called the "movie-in-the-brain." But those studies do not provide an answer to the puzzle yet, and I wish to make clear that I am not providing an answer either. When I attempted to elucidate feelings in Chapter Three, for example, I was trying to explain how they can be construed in a body with a brain, and why the construction of feelings is different, neurobiologically speaking, from the construction of other mental events. At the level of systems, I can explain the process up to the organization of neural patterns on the basis of which mental images will arise. But I fall short of suggesting, let alone explaining, how the last steps of the image-making process are carried out.[11]

The Construction of Reality

This perspective has important implications for how we conceive the world that surrounds us. The neural patterns and the corresponding mental images of the objects and events outside the brain are creations of the brain related to the reality that prompts their creation rather than passive mirror images reflecting that reality. For example, when you and I look at an external object, we form comparable images in our respective brains, and we can describe the object in very similar ways. That does not mean, however, that the image we see is a replica of the object. The image we see is based on changes that occurred in our organisms, in the body and in the brain, as the physical structure of that particular object interacts with the body. The ensemble of sensory detectors are located throughout our bodies and help construct neural patterns that map the comprehensive *interaction* of the organism with the object along its many dimensions. If you are watching and listening to a pianist play a certain piece, say Schubert's D. 960 sonata, the comprehensive interaction includes patterns that are visual, auditory, motor (related to the movements made in order to see and hear), and emotional. The emotional patterns result from the reaction to the person playing, to how the music is being played, and to characteristics of the music itself.

The neural patterns corresponding to the above scene are constructed according to the brain's own rules, and are achieved for a brief period of time in the multiple sensory and motor regions of the brain. The building of those neural patterns is based on the momentary selection of neurons and circuits engaged by the interaction. In other words, the building blocks exist within the brain, available to be picked up—selected—and assembled in a particular arrangement. Imagine a room dedicated to Lego play, filled with every Lego piece conceivable, and you get part of the picture.[12] You could construct anything you fancied, as does the brain because it has component pieces for every sensory modality.

The images we have in our minds, then, are the result of interactions between each of us and objects that engaged our organisms, as mapped in neural patterns constructed according to the organism's design. It should be noted that this does not deny the reality of the objects. The objects are real. Nor does it deny the reality of the interactions between object and organism. And, of course, the images are real too. And yet, the images we experience are brain constructions *prompted* by an object, rather than mirror reflections of the object. There is no picture of the object being transferred optically from the retina to the visual cortex. The optics stop at the retina. Beyond that there are physical transformations that occur in continuity from the retina to the cerebral cortex. Likewise, the sounds you hear are not trumpeted from the cochlea to the auditory cortex by some megaphone, although physical transformations do travel from one to the other, in a metaphorical sense. There is a set of *correspondences*, which has been achieved in the long history of evolution, between the physical characteristics of objects independent of us and the menu of possible responses of the organism. (The relation between the physical characteristics of the external object, and the a priori components that the brain selects to construct a representation are an important issue to explore in the future.) The neural pattern attributed to a certain object is constructed according to the menu of correspondences by selecting and assembling the appropriate tokens. We are so biologically similar among ourselves, however, that we construct similar neural patterns of the same thing. It should not be surprising that similar images arise out of those similar neural patterns. That is why we can accept, without protest, the conventional idea that each of us has formed in our minds the reflected picture of some particular thing. In reality we did not.

Seeing Things

How do we know that mental images and neural patterns are closely related and that the former come from the latter? We began to know of this close relation from the studies of David Hubel and Torsten Wiesel. They showed that an experimental animal (a monkey) looking at a straight line, curved line, or lines positioned at varied angles will form distinct patterns of neural activity in its visual cortex.[13] They also related the appearance of the distinct patterns to the microscopic anatomy of the visual cortex, thereby discovering the modular components with which you can construct a certain form. Further evidence came from an experiment by Roger Tootell in which an experimental animal (also a monkey) confronted a visual stimulus, for example, a cross, and a directly correspondent pattern could be identified in a specific layer of the animal's visual cortex—layer 4B of the primary visual cortex, also known as Brodmann's area 17 or area V1.[14] This demonstration brings together the key aspects of the process: the external stimulus, which we as observers can see as a mental image and which we can reasonably assume the experimental animal can see as a mental image as well; and the neural pattern that is clearly caused to happen as a result of seeing the stimulus. The experiment demonstrates multiple correspondences—the visual stimulus; the image we form related to it and which the animal presumably forms as well; and the neural pattern in the animal's brain. In that neural pattern *we*, as observers, can see a correspondence with our own image pattern, and, by extension, with the animals' image pattern.

We get an inkling of how this remarkable body mechanism may have evolved when we consider the visual devices available to a very simple creature, a variety of marine invertebrate known as *Ophiocoma wendtii*. *O. wendtii* is a brittle star, capable of fleeing rapidly and effectively from an approaching predator, and taking refuge in rocky caves and crevices in the vicinity. Given that the animal's external skeleton is made of hard calcium, that it lacks eyes, and that its nervous system is quite primitive, these evasive behaviors have long been a mystery. It turns out, however, that a good part of the animal's body is made of tiny calcium lenses that behave very much like an eye. The lenses focus incoming light into a small area beneath each lens where a bundle of nerves can then become active as a result. A predator's pattern can be mapped this way, and so can the pattern formed by a nearby crevice that can serve as hideaway. Processing of the predator patterns leads to nerve activation and to appropriate motor responses toward the protective crevice.[15] By no means am I suggesting that this creature thinks, although we can be certain it acts, and that it does so on the basis of freshly formed neural patterns. I am not even inclined to believe that, in such a simple nervous system, those neural patterns become mental images necessarily. I am simply using these facts to illustrate the genealogies of body-to-nervous system signaling on the basis of which body-to-mind influences can be understood. The human eye and its retina do something quite similar to the lenses of O. wendtii. But the eye's mechanism is far more complex in the variety of physical impingements that can be mapped, in the richness of subsequent mappings that can be formed, and in the wealth of actions that can be taken as a consequence. The essence, however, is the same: a specialized part of the body is modified and the result of the modification is transferred to the central nervous system.

A related finding that has become clear recently regards the presence of a special class of retinal cells that respond to light and influence the operation of a nucleus in the hypothalamus— the suprachiasmatic nucleus—known to regulate the day-night cycles and respective sleep patterns. That the rods and cones that form the front layer of the retina respond to light has long been known, and their responses are essential for vision. The intriguing new finding is that the influence of light on the hypothalamus is not mediated by the rods and cones; after destruction of the rods and cones, the light continues to pace the day-night cycle. A set of cells in the next layer—the retinal ganglion cell layer—seems to do this job. Moreover, the set of retinal ganglion cells that carries out this job is quite distinct—the retinal ganglion cells that receive signals from the rods and cones are *not* involved in the operation. This subset is present, it seems, for this particular operation alone and not at all to help with vision.[16] Directly or indirectly, activity in these cells exerts an influence on the mind. For example, turning on sleep diminishes attention and eventually suspends consciousness; background emotions and the related moods also are heavily influenced by overall exposure to light in terms of hours and intensity. Once again, a change in the state of the body—a specialized part of the body—translates itself in mental changes. Of great interest is the fact that the cells in question—unlike those that help with vision—are not interested in precisely where the light falls. Slowly and calmly, they respond like the light meters we use in photography to the overall luminance and the radiant light diffused inside the eye. It is tempting to see these cells as part of older and less sophisticated body sensors and preoccupied with overall conditions—namely the amount of ambient light surrounding an entire organism—rather than with the detailed shaping of light caused by external objects. In this sense they resemble the lenses of *O. wendtii,* and the whole body sensitivity that can be found in simpler organisms whose bodies are not equipped with specialized sensing regions.[17]

In the past twenty years, neuroscience has revealed in great detail how the brain processes various aspects of vision, not just shape but color and movement too.[18] Progress is also being made in the understanding of audition, touch, and smell, and at long last there is a renewed interest in the understanding of the internal senses—pain, temperature, and the like. It is fair to say, however, that we have barely begun to unravel the fine details of these systems.

About the Origins of the Mind

The two kinds of body image we have been considering, from the flesh and from special sensory probes, can be manipulated in our minds and used to represent spatial relations and temporal relations among objects. This allows us to represent events involving those objects. Are the images in our mind body images in the sense discussed above? Well, not exactly. Thanks to our creative imagination we can invent additional images to symbolize objects and events and to represent abstractions. For example, we can fragment the foundational images from the body we discussed earlier, and recombine the parts. Any object and event can be symbolized by some kind of invented, imageable sign, such as a number or a word, and such signs can be combined in equations and sentences. The invented, imageable signs can represent abstract entities and events just as well as concrete ones.

The influence of the body in the organization of the mind also can be detected in the metaphors our cognitive systems have developed to describe events and qualities in the world. Many of those metaphors are based on the work of our own imagination regarding the typical activities and experiences of the human body, such as postures, attitudes, direction of movement, feelings, and so on. For example, the ideas of happiness, health, life, and goodness are associated with "up" by word and gesture. Sadness, illness, death, and evil are associated with "down." The future is associated with "ahead." Mark Johnson and George Lakoff have explained quite persuasively how the categorization of certain body actions and postures has led to certain schemas that eventually are denoted by a gesture or word.[19]

At this point, I should inject another important qualification to this discussion. When we say that the mind is built from ideas that are, in one way or another, brain representations of the body, it is easy to conceive of the brain as a blank slate that begins its day clean, ready to be inscribed with signals from the body. But nothing could be further from the truth. The brain does not begin its day as a *tabula rasa*. The brain is imbued at the start of life with knowledge regarding how the organism should be managed, namely how the life process should be run and how a variety of events in the external environment should be handled. Many mapping sites and connections are present at birth; for example, we know that newborn monkeys have neurons in their cerebral cortex ready to detect lines in a certain orientation.[20] In brief, the brain brings along innate knowledge and automated know-how, predetermining many ideas of the body. The consequence of this knowledge and know-how is that many of the body signals destined to become ideas, in the manner we have discussed so far, happen to be engendered by the brain. The brain commands the body to assume a certain state and behave in a certain way, and the ideas are based on those body states and body behaviors. The prime example of this arrangement concerns drives and emotions. As we have seen, there is nothing free or random about drives and emotions. They are highly specific and evolutionarily preserved repertoires of behaviors whose execution the brain faithfully calls into duty, in certain circumstances. When the sources of energy in the body become low, the brain detects the decline and triggers a state of hunger, the drive that will lead to the correction of the unbalance. The idea of hunger arises from the representation of the body changes induced by the deployment of this drive.

To say that a good many ideas of the body are a consequence of the brain having placed the body in a particular state means that some of the ideas of the body that come to constitute the basics of the mind are highly constrained by prior design of the brain and the organism's overall needs. They are ideas of body actions, yet those body actions were first dreamed by a brain that commanded them to occur in a corresponding body.

The arrangement underscores the "body-mindedness" of the mind. The mind exists because there is a body to furnish it with contents. On the other hand, the mind ends up performing practical and useful tasks for the body—controlling the execution of automated responses in relation to the correct target; anticipating and planning novel responses; creating all sorts of circumstances and objects that are beneficial to the body's survival. The images that flow in the mind are reflections of the interaction between the organism and the environment, reflections of how the brain's reaction to the environment affects the body, reflections of how the body's adjustments are faring in the unfolding life state.

Someone might argue that since the brain provides the most immediate substrates of the mind—neural maps—the critical component to consider in the mind-body problem is the body's brain, not the body-proper. What do we gain by considering the mind in the perspective of the body, as opposed to considering the mind in the perspective of just the brain? The answer is that we gain a rationale for the mind that we would not discover if we considered the mind only in the perspective of the brain. The mind exists for the body, is engaged in telling the story of the body's multifarious events, and uses that story to optimize the life of the organism. Much as I dislike sentences that require laborious parsing, I am tempted to offer one as a summary of my view: The brain's body-furnished, body-minded mind is a servant of the whole body.

But now come several delicate questions. Why do we need a "mind level" of brain operations as opposed to a "neural-map level" alone as currently described by the tools of neuroscience? Why would a neural-map level, with activities neither mental nor conscious, be less efficient at managing the life process than the conscious-mind level? In even clearer terms, and in keeping with my line of thinking: Why do we need the neurobiological level of operations that also includes what we call mind and consciousness?

We can answer some of these questions and we can speculate about the others. For example, in the absence of consciousness in the comprehensive sense of the term—a process that includes both the movie-in-the-brain and the sense of self—we know for certain that life cannot be properly managed. Even temporary suspensions of consciousness entail an inefficient management of life. In effect, even the mere suspension of the self-component of consciousness entails a disruption of life management and returns a human being to a state of dependence comparable to that of a toddler. (This occurs in situations such as akinetic mutism.) It is certain that the conscious-mind level is a necessity for survival.

But what exactly is the indispensable contribution that the conscious-mind level of biology brings to the organism? Here the answers are speculative. As suggested in Chapter Four, perhaps the sheer complexity of sensory phenomena at the mental level permits easier integration across modalities, e.g., visual with auditory, visual and auditory with tactile, etc. In addition, the mental level also would permit the integration of actual images of every sensory stripe with pertinent images recalled from memory. Moreover, these abundant integrations would prove fertile ground for the image manipulation required for problem-solving and creativity in general. The answer, then, is that mental images would allow an ease of manipulation of information that the neural-map level (as described so far) would not permit. It is likely that in order to enable these new functions, the mind level of operations possesses biological specifications in addition to those present at the "current" neural-map level. That does not mean, however, that the mind level of biological operations is based on a different substance, in the Cartesian sense. The complex, highly

integrated images of mental process still can be conceived as biological and physical.

Now, we should consider what the sense of self brings to the
process. The answer is an *orientation*. The sense of self intro-
duces, within the mental level of processing, the notion that all
the current activities represented in brain and mind pertain to
a single organism whose auto-preservation needs are the basic
cause of most events currently represented. The sense of self ori-
ents the mental planning process toward the satisfaction of those
needs. That orientation is only possible because feelings are integral to the cluster of operations that constitutes the sense of
self, and because feelings are continuously generating, within the
mind, a *concern* for the organism.

In brief, without mental images, the organism would not be
able to perform in timely fashion the large-scale integration of information critical for survival, not to mention well-being. Moreover, without a sense of self and without the feelings that integrate
it, such large-scale mental integrations of information would not
be oriented to the problems of life, namely, survival and the
achievement of well-being.

This view of mind does not fill the knowledge gap I alluded to
earlier when I wrote that the current neuroscientific descriptions
of neural-map activities do not provide enough detail to tell us
about the biophysical composition of mental images. The gap
is acknowledged, as is the hope that it can be bridged in the
future.[21]

For the time being it is not unreasonable to conceive of the
mind as emerging from the cooperation of many brain regions.
This occurs when the sheer accumulation of details regarding the
state of the body that is mapped in those regions reaches a "critical pitch." The knowledge gap we now recognize may turn out to
be little more than a discontinuity in the complexity of the accu-

mulated detail, and in the complexity of the interactions of the brain regions involved in the mapping.

Body, Mind, and Spinoza

This is the time to return to Spinoza and to consider the possible meaning of what he wrote on body and mind. Whatever interpretation we favor for the pronouncements he made on the issue, we can be certain Spinoza was changing the perspective he inherited from Descartes when he said, in *The Ethics*, Part I, that thought and extension, while distinguishable, are nonetheless attributes of the same substance, God or Nature. The reference to a single substance serves the purpose of claiming mind as inseparable from body, both created, somehow, from the same cloth. The reference to the two attributes, mind and body, acknowledged the distinction of two kinds of phenomena, a formulation that preserved an entirely sensible "aspect" dualism, but rejected substance dualism. By placing thought and extension on equal footing, and by tying both to a single substance, Spinoza wished to overcome a problem that Descartes faced and failed to solve: the presence of two substances and the need to integrate them. On the face of it, Spinoza's solution no longer required mind and body to integrate or interact; mind and body would spring in parallel from the same substance, fully and mutually mimicking each other in their different manifestations. In a strict sense, the mind did not cause the body and the body did not cause the mind.

Were Spinoza's contribution on this issue limited to the above formulation, one would have to grant him that progress had been made. One would have to note, however, that by relating mind and body to a closed, single-substance box, he turned his back on the attempt to explain how the bodily and mental manifestations of substance ever arose. A fair-minded critic would add that at least Descartes was trying, while Spinoza merely circumvented the problem. But perhaps the fair-minded critic would not be accurate. In my interpretation, Spinoza was making a bold attempt at penetrating the mystery. I venture, and am ready to admit I may be wrong, that based on his statements in Part II of *The Ethics*, Spinoza may have intuited the general anatomical and functional arrangement that the body must assume for the mind to occur together with it, or, more precisely with and within it. Let me explain why I think so.

We should begin by reviewing Spinoza's notions of body and mind. Spinoza's notion of the human body is conventional. This is how he describes the body in *The Ethics*, Part I: "a definite quantity, so long, so broad, so deep, bounded by a certain shape." Using Spinozian phrasing, my own description would be "a certain amount of substance, fenced in." And since Spinoza's substance is Nature, I would say, "a body is a lump of Nature, properly fenced in by the boundary of the skin."

For details on Spinoza's conception of the body we must turn to the set of six postulates in Part II of the *Ethics*. They are:

I. *The human body is composed of a number of individual parts, of diverse nature, each one of which is in itself extremely complex.*

II. *Of the individual parts composing the human body some are fluid, some soft, some hard.*

III. *The individual parts composing the human body, and consequently the human body itself, are affected in a variety of ways by external bodies.*

IV. *The human body stands in need for its preservation of a number of other bodies, by which it is continually, so to speak, regenerated.*

V. *When the fluid part of the human body is determined by an external body to impinge often on another soft part, it changes*

*the surface of the latter, and, as it were, leaves the impression
thereupon of the external body which impels it.*

VI. *The human body can move external bodies, and arrange
them in a variety of ways.*

The dynamic image Spinoza conveys is quite sophisticated,
especially when we remember that this was written in the middle
of the seventeenth century and that the ink was still fresh on the
first treatises of anatomy. There were many parts to this complex
body thing. They were perishable and had to be renewed. They
could be deformed by contact with other bodies. He stopped
short of saying that the deformations could be conveyed by
nerves to the brain, although I would not put it past him to have
thought so.

The real breakthrough, as I see it, regards Spinoza's notion of
the human mind, which he defines transparently as consisting of
the idea of the human body. Spinoza uses "idea" as a synonym for
image or mental representation or component of thought. He
calls it "a mental conception which is formed by the mind of a
thinking entity." (Elsewhere, however, Spinoza uses idea to signify an elaboration on images, as a product of the intellect rather
than of plain imagination.)

Consider Spinoza's exact words: *"The object of the idea constituting the human Mind is the Body,"* which appears in Proposition
13 of Part II of *The Ethics.*[22] The statement is reworded and elabo-
rated in other propositions. For example, within the proof of
Proposition 19, Spinoza says, "The human mind is the very idea or
knowledge of the human body." In Proposition 23, he states, "The
Mind does not have the capacity to perceive ... except in so far as it
perceives the ideas of the modifications (affections) of the body."

Consider, moreover, the following relevant passages, all from
Part II of *The Ethics:*

(a) ... *The object of the idea constituting the human mind is the body, and the body as it actually exists... Wherefore the object of our mind is the body as it exists, and nothing else...* (From the proof following Proposition 13).

(b) *We thus comprehend, not only that the human mind is united to the body, but also the nature of the union between mind and body* (and)

(c) ... *in order to determine, wherein the human mind differs from other things, and wherein it surpasses them, it is necessary for us to know the nature of its object, that is, of the human body. What this nature is, I am not able here to explain, nor is it necessary for the proof of what I advance, that I should do so. I will only say generally, that in proportion as any given body is more fitted than others for doing many actions or receiving many impressions at once, so also is the mind, of which it is the object, more fitted than others for forming many simultaneous perceptions...* (From the note following Proposition 13).

The latter concept is worded in ringing form in Proposition 15: "*The human mind is capable of perceiving a great number of things, and is so in proportion as its body is capable of receiving a great number of impressions.*"

Perhaps most importantly, consider Proposition 26: "*The human Mind does not perceive any external body as actually existing except through the ideas of the modification (affections) of its own body.*"

Spinoza is not merely saying that mind springs fully formed from substance on equal footing with body. He is assuming a mechanism whereby the equal footing can be realized. The mechanism has a strategy: Events in the body are represented as ideas in the mind. There are representational "correspondences," and they go in one direction—from body to mind. The means to achieve the representational correspondences are contained in the substance. The statements in which Spinoza finds ideas "proportional" to "modifications of the body," in terms of both quantity and intensity, are especially intriguing. The notion of "proportion" conjures up "correspondence" and even "mapping." I suspect he is referring to some sort of structure-preserving isomorphy. Equally stirring is his notion that the mind cannot perceive an external body as existing, except through the modifications of its own body. He is in effect specifying a set of functional dependencies: He is stating that the idea of an object in a given mind cannot occur without the existence of the body; or without the occurrence of certain modifications on that body as caused by the object. No body, never mind.

Spinoza does not venture beyond his knowledge and thus could not say that the means to establish ideas of the body include chemical and neural pathways and the brain itself. Of necessity Spinoza knew very little about the brain and about the means for body and brain to signal mutually. Spinoza was cautious to claim ignorance of the body's anatomical and physiological detail, including the part of the body called brain. He carefully avoids mentioning the brain when he discusses mind and body, although we can be certain from statements elsewhere that he saw brain and mind as closely associated. For example, in the discussion that closes Part I of *The Ethics,* Spinoza says that "everyone judges of things according to the state of his brain." In the same discussion he interprets the proverb, "brains differ as completely as palates," to mean that "men judge of things according to their mental disposition." Be that as it may, now we can fill in the brain details and venture to say for him what he obviously could not.

From my current perspective, to say that our mind is made up of ideas of one's body is equivalent to saying that our mind is made up of images, representations, or thoughts of our own parts of our own body in spontaneous action or in the process of modifications caused by objects in the environment. The statement departs radically from traditional wisdom and may sound implausible at first glance. We usually regard our mind as populated by images or thoughts of objects, actions, and abstract relations, mostly related to the outside world rather than to our bodies. But the statement is plausible, considering the evidence I presented on the processes of emotion and feeling in Chapters Two and Three, and on the neurophysiology evidence discussed in this chapter. The mind is filled with images from the flesh and images from the body's special sensory probes. On the basis of findings from modern neurobiology, not only can we say that images arise in the brain, but we can venture that a vast proportion of the images that ever arise in the brain are shaped by signals from the body-proper.

I regard the Spinoza of *The Ethics*, Part I, where he addressed the issues of mind and body in general, as the consummate philosopher dealing with the whole universe. In Part II, however, Spinoza was concerned with a local problem, and I suspect he was intuiting a solution he could not specify. The result of this double perspective ranges from a latent tension to a seeming clash, the sort of conflict that permeates the *Ethics*. After all, the equal footing of mind and body only works in the general description. Once Spinoza goes deep inside the unspecified mechanism there are preferred directions of operation, from body to mind when we perceive, and from mind to body when we decide to speak and do so.

Spinoza does not hesitate to privilege body or mind in certain circumstances. In most of the propositions discussed thus far, the body quietly wins, of course. But in Proposition 22 (*The Ethics*, Part II), Spinoza does privilege the mind: *"The human mind perceives not only the modifications of the body, but also the ideas of such modifications."* Which really means that once you form an idea of a certain object, you can form an idea of the idea, and an idea of the idea *of the idea*, and so forth. All of this idea formation occurs on the mind side of substance, which, in the current perspective, can be largely identified with the brain-mind sector of the organism.

The notion of "ideas of ideas" is important on many counts. For example, it opens the way for representing relationships and creating symbols. Just as importantly, it opens a way for creating an idea of self. I have suggested that the most basic kind of self is an idea, a second-order idea. Why second order? Because it is based on two first-order ideas—one being the idea of the object that we are perceiving; the other, the idea of our body as it is modified by the perception of the object. The second-order idea of self is the idea of the relationship between the two other ideas—object perceived *and* body modified by perception.

This second-order idea I call self is inserted in the flow of ideas in the mind, and it offers the mind a fragment of newly created knowledge: the knowledge that our body is engaged in interacting with an object. I believe such a mechanism is critical for the generation of consciousness, in the comprehensive sense of the term, and I have hypothesized processes that would permit the implementation of this mechanism in the brain.[23] We have a conscious mind when the flow of images that describes objects and events in varied sensory modalities—the movie-in-the-brain—is accompanied by the images of the self I just described. A conscious mind is a plain mind process that is being informed of its simultaneous and ongoing relationships to objects and to the organism that harbors it. Once again, it is intriguing that Spinoza made room in his thinking for an operation as simple and as interesting as making ideas of ideas.

Spinoza had no patience for arguments from ignorance, of the kind we often encounter when someone declares the unlikelihood of mind arising from biological tissue because "it is impossible to imagine." He was clear about the fact:

> ... No one has hitherto laid down the limits to the pow-
> ers of the body, that is, no one has as yet been taught by ex-
> perience what the body can accomplish solely by the laws of
> nature, in so far as she is regarded as extension. No one
> hitherto has gained such an accurate knowledge of the bod-
> ily mechanism, that he can explain all its functions ... No
> one knows how or by what means the mind moves the body,
> nor how many various degrees of motion it can impart to
> the body, nor how quickly it can move it. Thus, when men
> say that this or that physical action has its origin in the
> mind, which latter has dominion over the body, they are
> using words without meaning, or are confessing in specious
> phraseology that they are ignorant of the cause of the said
> action ... [24]

Here I suspect that Spinoza is referring to the body in a comprehensive way, body-proper *and* brain. Perhaps he was not only
undermining the traditional notion that the body would arise
from the mind, but also preparing the stage for discoveries that
would support the opposite notion.[25]

Others may disagree with my interpretation. For example, it
could be argued that my reading of Spinoza would be undermined by Spinoza's notion that the mind is eternal. The objection
would not be valid, however. At numerous junctions in *The Ethics*,
namely in Part V, Spinoza defines eternity as the existence of eternal truth, the essence of a thing, rather than a continuance over
time. The eternal essence of the mind is not to be confused with
immortality. In Spinoza's thinking the essence of our minds existed before our minds ever were, and persists after our minds
perish with our bodies. Minds are *both* mortal *and* eternal. Be-

sides, elsewhere in *The Ethics* and in the *Tractatus*, Spinoza declares the mind perishable with the body. In fact, his denial of the immortality of the mind, a feature of his thinking from his early twenties, may have been a principal reason for his expulsion from his religious community.[26]

What is Spinoza's insight then? That mind and body are parallel and mutually correlated processes, mimicking each other at every crossroad, as two faces of the same thing. That deep inside these parallel phenomena there is a mechanism for representing body events in the mind. That in spite of the equal footing of mind and body, as far as they are manifest to the percipient, there is an asymmetry in the mechanism underlying these phenomena. He suggested that the body shapes the mind's contents more so than the mind shapes the body's, although mind processes are mirrored in body processes to a considerable extent. On the other hand, the ideas in the mind can double up on each other, something that bodies cannot do. If my interpretation of Spinoza's statements is even faintly correct, his insight was revolutionary for its time but it had no impact in science. A tree fell silently in the forest and no one was there to serve witness. The theoretical implications of these notions have not been digested either as Spinozian insight or as independently established fact.

Closing with Dr. Tulp

I ended my Huygens Lecture by showing a reproduction of Rembrandt's *Anatomy Lesson of Dr. Tulp,* which hangs close by in the Mauritshuis. It was not the first time I used Dr. Tulp in reference to the mind-body problem, but for once the place and the topic were in perfect tune.

On the face of it, Rembrandt's painting celebrates Dr. Tulp's fame as a physician and scientist on the occasion of a particular anatomy lesson in January 1632. The Guild of Surgeons wished to honor Dr. Tulp with a painting and no better theme could be found than a theatrical anatomical dissection, a public and paid event that attracted the curiosity of the educated and the wealthy. But the painting also celebrates a new era in the study of the body and its functions, chronicled in the writings of William Harvey and Descartes, who is presumed to have been in the audience on that day. Harvey's discoveries on the circulation of the blood are of the same vintage, the post-Vesalius era of fine scalpels, lenses, and microscopes that could dissect and amplify the fine physical structure of the human body. The work announced the Dutch interest in studying and depicting nature—all the way into the human body, underneath its skin—and was a good emblem of the rise of science that marked this age.

Perhaps more importantly, Rembrandt's painting also reminds us of the puzzlement that the new anatomical discoveries produced in the discoverers. Dr. Tulp's right hand holds the tendons with which the cadaver's left hand once flexed its fingers, while Dr. Tulp's own left hand demonstrates the motion those tendons would accomplish. The mystery behind the action is revealed for all to see. It is not a hydraulic or pneumatic pump device, although it might have been, of course, and therein lies the beauty of the moment captured on the canvas: The movement of a hand is achieved by muscular contraction and by the related pulling of tendons attached to bony parts, rather than in some other way. Dr. Tulp verifies what *is* and separates what is from what *might be*. Conjecture gives way to fact.

The spectacle of a mystery revealed, however, is disquieting for some, and that is the least we can read in Dr. Tulp's look. Dr. Tulp does not face the viewer, nor does he look at what he is doing, nor does he glance at his colleagues. He stares leftward into a distance beyond the confines of the frame and, if historian Simon Schama is correct, beyond the confines of the room. Schama suggests that Dr. Tulp is looking at the Creator himself. The interpretation accords well with the fact that Tulp was a devout Calvinist and with these verses written by Caspar Barleus a few years later after the painting gained renown: "Listener, learn yourself and while you proceed through the parts, believe that, even in the smallest, God lies hid."[27] I see Barleus's words as a response to the unease of the discovery, the unease that would have been produced by the inevitable, subsequent thought: If we can explain this about our nature, what can we not explain? Why can we not explain everything else that happens in the body, including, perhaps, the mind? Will we be able to discover how one's thoughts can *will* a hand to move? Afraid of his own thoughts, Barleus wished to calm the public, or the deity, or both, by saying that, although they are trespassing backstage and discovering how the tricks are done, by no means are they less reverent for the work of the Creator. The intended meaning of Dr. Tulp's facial expression is impossible to decipher, of course, and sometimes when I stand in front of the picture I think that he is simply telling the viewer: "Look what I've done!" Whatever the precise meaning, Rembrandt or Tulp, perhaps both, wanted us to know that no one took in stride what was happening in the *Theatrum Anatomicum*.[28]

Barleus's pious reassurance was indeed necessary as an antidote against what Descartes probably was thinking on those days regarding mind and body, and especially so against what Spinoza would be thinking and writing on this issue within the next two decades. And it is fascinating to realize—showing once again how words can lie—that if you would take Barleus's admonition out of context and offer it as Spinoza's, the meaning would be entirely different. Looking at Rembrandt's masterpiece, Spinoza could perfectly well have said, that *his* God was in every inch and every motion of the dissected body, yet he would have signified something else.

CHAPTER 6

A Visit to Spinoza

Rijnsburg, July 6, 2000

I am sitting in the small garden behind Spinoza's house. The sun is out, the air is actually warm, and the silence is almost complete. Few people drive or walk in the Spinozalaan. Only a black cat is moving, appearing serene and absorbed in his preparations for a heavenly, philosophical summer day.

I am looking at the same sky Spinoza must have looked at if he ever walked out of one of his rooms and sat in this same spot. And if he did not, on such a day as this, the sun would walk in and come to his desk, a most welcome event in this climate. This is a nice place, less confining than the house in The Hague but still too modest a perch for someone who was observing the entire universe.

How does one become Spinoza, I ask myself? Or, to phrase it differently, how can we explain his strangeness? Here is a man who firmly disagreed with the leading philosopher of his time, publicly battled organized religion and was expelled from his own, rejected the way of life of his contemporaries, and set goals for his own way of life that some considered saintly and many considered foolish. Was Spinoza the social aberration he has been made out to be? Or is he understandable in terms of the culture of his time and place? Can his behavior be explained by events in his personal life? I am intrigued by these questions. Leaving aside the fool-hardiness of attempting to account for anyone's life satisfactorily, I believe some tentative answers are possible.

The Age

In spite of his originality, Spinoza does not stand alone in his historical time. He rose in the middle of the century of genius, the seventeenth, the period during which the foundations of the modern world were laid down. Spinoza was a radical, but so was Galileo when he confirmed and endorsed Copernicus just about the time Spinoza was born. This was a century that began with Giordano Bruno being burned at the stake and with the first per-formances of Shakespeare's mature version of *Hamlet* (1601). By 1605 the world had been treated to Francis Bacon's *Advancement of Learning*, Shakespeare's *Lear*, and Miguel de Cervantes's *Don Quixote*. Hamlet may well be the emblem of the entire age be-cause he traverses Shakespeare's longest play bewildered by human behavior and puzzled by the possible meaning of life and death. On the surface the plot may be about the failure to avenge a wronged father and kill an uncle who is less than kind. But the theme of the play is Hamlet's puzzlement, the disquietude of a man who knows more than those around him and yet not enough to quench his discomfort with the human condition. Hamlet is aware of the science of the day—physics and biology, such as they are; he goes to the University of Wittenberg, after all—and he knows about the intellectual dislocations brought about by Martin Luther and Jean Calvin. But because he cannot make sense of what he sees, he questions and kvetches at every possible turn. It's no coincidence that the word "question" appears more than a dozen times in *Hamlet* or that the play begins with a particular question: "Who's there?" Spinoza was born into the age of questioning, an era that might as well be known as the Hamlet age.

Spinoza also was born into the age of the observable fact, when the antecedents and the consequences of a given action began to be studied in experiments rather than debated from the comfort of an armchair. The human intellect already was in full command of a means to reason logically and creatively in the manner demonstrated by Euclid. However, to use Albert Einstein's words, "before mankind could be ripe for a science which takes in the whole of reality, a second fundamental truth was needed... all knowledge of reality starts from experience and ends in it."[1] Einstein singled out Galileo as the epitome of this attitude—he saw him as the "father of modern science altogether"—but Bacon was another leading proponent of the new approach. Both Galileo and Bacon advocated experimentation and proceeded by the gradual elimination of false explanations. And Galileo added something else: He believed the universe could be described in the language of mathematics, a notion that would provide a cornerstone for the emergence of modern science. Spinoza's birth coincided with the first flowering of science in the modern world.

The importance of measurement was established at this time and science became quantitative. Scientists now used the inductive method as a tool, and empirical verification became the foundation for thinking about the world. An open season was declared on ideas that did not accord with fact.

This epoch was so intellectually crowded that roughly about the time Spinoza was born Thomas Hobbes and Descartes were rising as philosophical figures and William Harvey was describing the circulation of the blood. Within Spinoza's own brief life, the world also would learn about the work of Blaise Pascal, Johannes Kepler, Huygens, Gottfried Leibniz, and Isaac Newton (who was born a mere ten years after Spinoza). As Alfred North Whitehead says aptly, "There simply was not time for the century to space out nicely its notable events concerning men of genius."[2]

Spinoza's general attitude toward the world was part of this new questioning ferment, and was rooted in some remarkable changes in the manner in which explanations were formulated and institutions assessed. But knowing where Spinoza fits in the grand historical scheme and discovering that his brilliance had company does not explain why Spinoza was the figure of the century whose work was banned most fiercely, so fiercely that there is hardly any reference to his ideas for decades on end, unless it is derogatory. Spinoza may have been no more radical than Galileo in his observations, but he was more contusive and even more uncompromising. He was the most intolerable kind of iconoclast. He threatened the edifice of organized religion at its foundations, at once fearlessly and modestly. By extension he threatened the political structures closely associated with religion. Predictably, the monarchies of the time sensed the danger, and so did his own Dutch provinces, the most tolerant state of this era. What sort of life story can possibly help account for the development of such a cast of mind?

The Hague, 1670

When I try to understand Spinoza's life trajectory, I always return to The Hague and to his arrival at the Paviljoensgracht, during a brief calm in between storms, as a pivotal viewpoint to explain the befores, the afters, and the becauses. Spinoza was thirty-eight when he arrived in The Hague, alone, as was his custom. He brought a bookcase with his library, a desk, a bed, and his lens-making equipment. He would complete *The Ethics* in the two rooms he rented in the Paviljoensgracht, work daily at the manufacturing of lenses, receive hundreds of visitors and rarely travel any significant distance. He would go to Utrecht once and to Amsterdam many times, neither being much farther than thirty miles from The Hague, but he never went further than that. One thinks of Immanuel Kant, another distinguished loner a century later who managed to beat Spinoza's record: He spent an entire lifetime in Königsberg and is said to have ventured out of the city only once. Beyond the aversion to travel and the intellectual caliber, there is little resemblance between the two men. Kant wished to combat the perils of passion with dispassionate reason; Spinoza wished to combat a dangerous passion with an irresistible emotion. The rationality Spinoza craved required emotion as an engine. The two men were not alike in manner, either, as far as I can picture them. Kant, at least late Kant, was tense and formal, the epitome of polite circumspection. A bit of a dry stick. Spinoza was amiable and relaxed, albeit courtly and ceremonial in gesture. Late Spinoza—if we can talk about late when someone reaches forty—was kind, almost sweet, in spite of his quick wit and sharp tongue.

For a few months before moving to the Paviljoensgracht Spinoza had rented rooms just around the corner in the Stilleverkade. But the rent was too high, or so he thought, and he did not stay long. Before the Stilleverkade he had lived for seven years in

Voorburg, a small suburb due east of The Hague; and before that, he had spent two years in Rijnsburg, a town close to Leiden, midway between Amsterdam and The Hague. From the time he left the family home to the time he moved to Rijnsburg, Spinoza lived in varied places in Amsterdam or nearby. Sometimes he was the guest of friends, sometimes a boarder. He never owned a house and he never occupied more than a bedroom and a study.

Spinoza's frugality was self-imposed. Notwithstanding the ups and downs of his father's business, Spinoza was born into a wealthy family. His uncle Abraham was one of Amsterdam's wealthiest merchants, and Spinoza's mother had brought a large dowry into her marriage. Yet by his late twenties Spinoza had become indifferent to personal wealth and social status, although he continued to see nothing wrong with business profits. He simply did not find money and possessions rewarding, although he thought they could be, for others, and that the determination of how much wealth one should accumulate and how much spending was needed or appropriate rested with each individual. Let each be the judge.

He arrived at this attitude toward wealth and social status gradually and amid conflict. Spinoza appreciated the value of his education and knew that it would not have been possible without his family's financial and social position. Between his late adolescence and age twenty-four, he was a businessman and for a time he was in charge of the family firm. At that point, he certainly cared for money enough to take fellow Jews to the Dutch court when they did not pay their debts. This was a brazen act from the perspective of the community because any kind of conflict among Jews was to be resolved within the walls of the community and by its leaders. And when his father died leaving the firm with a considerable number of debts, Spinoza did not hesitate to make himself a warden of the Dutch court and to be named priority creditor of his inheritance. On the subject of money and possessions this last episode was a watershed. Spinoza renounced the inheritance altogether except for one item: his parents' bed. The *ledikant* would accompany him from place to place, and he would eventually die in it. I find the fixation on the *ledikant* fascinating, by the way. Of course, there were practical reasons to keep the bed, at least for some time. A *ledikant* is a canopied, four-poster bed with heavy curtains that can be drawn to transform it into a warm, isolated island. In Spinoza's time the *ledikant* was a sign of affluence. The common bed in Amsterdam houses was the *armoire* bed—literally, a bed inside a spacious wall closet whose doors could be open at night. But imagine holding on to the bed in which your parents conceived you, in which you played as an infant, and in which your parents died, and deciding to sleep there forever, live there, practically. Spinoza never had to dream about a long lost Rosebud because he never had to let go of it.

Midway through Spinoza's short life, historical circumstances had reduced the value and profitability of the family firm, although this was hardly a catastrophic collapse. There is little doubt that as a smart and enterprising businessman, Spinoza could have turned these failing fortunes around. But by then Spinoza had discovered thinking and writing as his sources of satisfaction, and needed little to support a life dedicated to them. On several occasions, Spinoza's friend Simon de Vries attempted to provide him with a stipend, but Spinoza never accepted. When the dying de Vries tried to make Spinoza his heir, Spinoza dissuaded him and insisted he would only accept a small annuity to help make ends meet, a sum of 500 florins. And when de Vries died and bequeathed the small pension they had agreed upon, Spinoza reduced the sum further and accepted only 300 florins. He told de Vries' baffled brother that the smaller amount would be more than sufficient. Later, he also refused a generous offer to become

a professor of philosophy at the University of Heidelberg—a position offered on Leibniz' recommendation—although the main
reason for the refusal probably had to do with the potential loss of
his intellectual freedom. Even so, declining the professorship certainly meant that he valued his thinking more than he did the
comforts that the Elector Palatine was making available in Heidelberg. Spinoza subsisted on his lens manufacturing work and,
after 1667, on the small pension from de Vries. The money was
enough to pay room and board; to buy paper, ink, glass, and tobacco; and to satisfy the doctor's bills. He required nothing else.

Amsterdam, 1632

Life was not always like this, for better or worse. Spinoza's father,
Miguel de Espinoza, was a prosperous Portuguese merchant, and
so had been Spinoza's paternal grandfather. When Spinoza was
born in 1632, Miguel was trading sugar, spices, dried fruit, and
Brazilian woods from his warehouse. He was a respected member
of the Jewish community, which numbered about 1,400 families,
almost exclusively of Sephardic Portuguese origin. He was a major
contributor to the Portuguese synagogue. On several occasions
he was a governor (a *parnas*) of the school and synagogue, and in
the last years of his life he was a member of the *mahamad,* the lay
governing group of the congregation. He was a close friend of
Rabbi Saul Levi Mortera, one of the most influential rabbis of this
period in Amsterdam. Uncle Abraham was a friend of Rabbi
Menassah ben Israel, another notable rabbi of the age. Like so
many Sephardic Jews, they had fled Portugal and the Inquisition,
first to Nantes in France and then to the low countries, establishing themselves in Amsterdam not long before Spinoza's birth. Spinoza's mother, Hana Deborah, also came from a prosperous
Sephardic Jewish family of Portuguese and Spanish lineage.

The Inquisition had been established much later in Portugal
than in Spain. In Portugal it began in 1536 and only gathered mo-

mentum after 1580. The long delay had given Portuguese Jews the opportunity to immigrate to Antwerp and later to Amsterdam, lands of greater promise than Northern Africa, Northern Italy, and Turkey, where Spanish Jews had immigrated a century earlier.

At the beginning of the seventeenth century, the Netherlands, and Amsterdam in particular, was indeed a promised land. Unlike virtually everywhere else in Europe, the social and political structure was marked by relative racial tolerance (extended to Jews, especially if they were Sephardic) and relative religious tolerance (extended gladly to Jews, but not so warmly to Catholics). The aristocracy was reasonably educated and benevolent. The House of Orange did have princes, but they held the post of stadholder, a president of sorts responsible to a council of the Dutch provinces. The Netherlands was a republic, and for a long period during Spinoza's life the stadholder was not the prince of Orange but rather an intelligent commoner. The Dutch introduced the makings of contemporary justice and modern capitalism. Commerce was respected. Money was supremely valued. The government created laws to permit citizens to buy and sell freely and to the best advantage. A large bourgeoisie flourished and devoted itself to the pursuit of property and a life of comfort. The more enlightened Calvinist leaders welcomed the contributions Portuguese Jewish merchants made to those pursuits.

In spite of the cultural uprooting, the Jewish community was culturally rich and financially affluent. There certainly were difficulties imposed by exile, internal religious tensions, and the need to comply with a host country. Yet the group probably was more close-knit than it would have been in Portugal dispersed over a far larger area and under the erratic shadow of the Inquisition. The Jews practiced their religion freely at home and in the synagogue. Business flourished and even managed to survive the economic downturns that followed multiple wars with Spain and Britain. It even was possible to use the mother tongue, Portuguese, without stigma, at home, at work, and in the synagogue.

There was no Jewish quarter in Amsterdam. Jews could reside anywhere they wished and could afford. Most affluent Jews chose to live around the Burgwaal, and this is where the Spinoza family lived, not far from where the Sephardic synagogue, which consolidated the three original Jewish communities of Amsterdam, eventually was built in the Houtgracht in 1639. (The impressive Portuguese synagogue that still stands today was erected nearby in 1675.) Many non-Jews had houses in the same area, and one of them was Rembrandt, who lived on the Breestraat in a house that still stands. There is no evidence that Rembrandt and Spinoza ever met, although from the overlap of dates (Rembrandt lived from 1606 to 1669; Spinoza from 1632 to 1677) they certainly could have. Rembrandt knew several members of the Jewish congregation, some of whom were avid art collectors. He painted several of them in portraits, street scenes, and in the synagogue, and illustrated a book by Menassah ben Israel, the most famous scholar of the time and eventually one of Spinoza's teachers. In turn, Rembrandt consulted ben Israel for the details of his painting of *Belshazzar's Feast*. It would be nice to discover that Rembrandt painted Spinoza's portrait, but there is no sign that he did. Legend has it that Rembrandt did use Spinoza's likeness in his painting *Saul and David*, which he created around the time Spinoza was expelled from the synagogue. The picture depicts David playing the harp for Saul (and is entirely different from Rembrandt's other painting on the subject, *David Playing the Harp for Saul*). David's frame and features could indeed be Spinoza's. More importantly, Spinoza could be reconceived as David—small but unexpectedly strong, capable of destroying Goliath and displeasing Saul, capable of being King himself.[3]

The limits imposed by the Protestant Dutch were few and clear. The Dutch had targeted Catholics as enemies, especially the Spanish Catholics with their demonic and bellicose expansionist plans. The Jews also considered Catholics enemies, especially the Spanish Catholics who, not content with creating a ferocious Inquisition, pressured the Portuguese to create their own. Under these circumstances the Jews and the Dutch were natural friends. Besides, the business of the Dutch was business and the Portuguese Jews brought good business to the Dutch provinces. The Jews controlled an extensive network of commercial and banking connections in the Iberian Peninsula, Africa, and Brazil, second to none. Descartes would say of Amsterdam that everyone but he was so engaged in trade and so mindful of his own profit that one could live an entire life there without ever being noticed. (That was wishful thinking and almost true, although Descartes hardly escaped attention.) When Spinoza was growing up, Jews accounted for about 10 percent of the members on the Amsterdam Stock Exchange and were vital for a number of missions having to do with arms sales and international banking. By 1672, Amsterdam's Jewish community had grown to about 7,500 members. It accounted for 13 percent of the bankers but less than four percent of the population. (Simon Schama points out that the prosperity of the Jewish community in Amsterdam probably is due to the fact that they were a significant but nondominant part of the city's life, banking included.4) It is hardly surprising that the Dutch were supportive of the Jews. As long as they did not try to convert the Protestants to the Jewish faith, or marry them, they were free to practice their religion and teach that religion to their children.

Never mind how welcoming Amsterdam was, one cannot imagine Spinoza's young life without the shadow of exile. The language was a daily reminder. Spinoza learned Dutch and Hebrew, and later Latin, but he spoke Portuguese at home, and either Portuguese or Castilian Spanish at school. His father always spoke Portuguese at work and at home. All transactions were recorded in Portuguese; Dutch was used only to deal with Dutch customers. Spinoza's mother never learned Dutch. Spinoza would lament that his mastery of Dutch and Latin never equaled that of Portuguese and Castilian. "I really wish I could write to you in the language in which I was brought up," he wrote to one of his correspondents.

Manners and dress were another reminder that, prosperity aside, this was exile rather than the homeland. The Sephardim were aristocratic in garb and bearing, cosmopolitan and worldly. Their ways reflected the life of the aristocratic businessmen in southern Europe—the word Sephardic refers to those who come from the cities of the south, known as Sepharad. Life in the Sepharad mixed work and socializing to a considerable degree, prompted perhaps by the milder climate. There was a concern for elegant and luxurious garments, and an ear attuned to news from the most faraway places, which arrived daily in the merchant ships calling at large ports such as Lisbon or Porto. The Dutch must have seemed too practical and hardworking by comparison.

Spinoza may have been destined, at first, for a career in business, but instead he became a brilliant student of Judaism, mentored by Rabbi Mortera and Rabbi ben Israel. Community leaders had brought these two Jewish scholars to Amsterdam in the hope of redressing the watering down of religious practices that followed centuries of sojourn in the Iberian Peninsula. The time was ripe for a revival of Jewish traditions now that the community was wealthy, geographically cohesive and religious practices no longer needed to be secret. The Jews formed a *nação,* the Portuguese word for nation, and Amsterdam would become a new Jerusalem in that nation. In this climate of rebirth and new hope, the prodigious intelligence of young Spinoza was properly cherished.

Spinoza proved to be a diligent and hardworking student. But the same diligence and inquisitiveness that made him an authority on the Talmud also made him question the foundations of the knowledge he so thoroughly absorbed. He was developing conceptions of human nature that eventually would diverge from that knowledge. The drift seems to have been gradual and the community probably did not notice until Spinoza had become a businessman around age eighteen or so. Even then there were no direct confrontations with the synagogue, mostly rumors, and Spinoza continued to be a member in good standing. The signs were clear, however. Spinoza had formed close friendships with several non-Jews, among them Simon de Vries, a wealthy business colleague whose family owned a splendid house on the Singel and an estate in Schiedam, near Amsterdam, and he was beginning to drift from the community. But worse was yet to come.

No later than age twenty and perhaps as early as eighteen, Spinoza enrolled in the school of Frans Van den Enden with the stated purpose of learning Latin. Van den Enden was a lapsed Catholic, a free thinker, a polyglot, and a polymath. He held both medical and law degrees and was knowledgeable in philosophy, politics, religion, music, the arts, you name it. Van den Enden's gargantuan appetite for life had not landed him in trouble, yet, but he created trouble for young Spinoza. At first quietly, then openly, first as an adolescent, then as a young man, Spinoza tasted life outside the community's paradise. He also spoke his mind and acted his mind. The community reacted with disappointment, then outrage.

By 1656, two years after his father died, the twenty-four-year-old Spinoza, now responsible for the family firm—"Bento y Gabriel

de Espinosa"—had continued to support the synagogue financially. Yet, liberated from any fear of embarrassing his father in
front of the community, he made no secret of his ideas regarding
the nature of human beings, God, and the practice of religion,
none of which accorded easily with Jewish teachings. His philosophy was taking shape and he talked freely about his ideas. No
amount of pleading from former mentors silenced his voice. No
appeals moved him. No bribes or threats changed his mind. A
murder attempt by a fellow Jew almost put an end to the community's embarrassment, although it is by no means certain that the
synagogue was behind the deed. The large cloak Spinoza was
wearing on the night he was to be stabbed kept the blade away
from his slim body. Spinoza lived to tell, undeterred, and kept
the cloak as a memento. Finally, as a last recourse, the synagogue decided to exclude him from the community altogether. In
1656 Spinoza was formally banished. Thus came to an end the
privileged life of he who was born Bento Spinoza, the name he
signed as a businessman, but was known by the community as
Baruch Spinoza. Thus began the twenty-one-year life of Benedictus Spinoza, the philosopher whose mature years were spent in
The Hague.

Ideas and Events

If Spinoza's small library is any indication, the new philosophy
and the new physics of his time were important influences in his
development. Books by Descartes and physicists were the most
frequent items in Spinoza's bookcase. Hobbes also was represented, as was Bacon. But Spinoza must have read prolifically in
his younger years, borrowing from his circle of well-read friends
books that we will never be able to trace. Without a doubt Spinoza
was acquainted with new methods for the evaluation of scientific
evidence, new facts from physics and medicine, and new ideas

presented by Descartes and Hobbes, perhaps the most read of modern thinkers during Spinoza's formative years. Spinoza was not a systematic experimenter—but then, neither was Bacon. Yet he had a grasp of empirical science from his readings and perhaps from his work in optics. He certainly knew how to evaluate facts. His achievements came from logical reflection on a considerable body of new scientific evidence and were complemented by a rich intuition.

Frans Van den Enden's school and the headmaster himself may have been critical catalysts in Spinoza's intellectual development. Van den Enden's circle was ideal for Spinoza to discuss ideas that obviously had been simmering in his young mind and that needed some open, if limited, debate to mature. Van den Enden ran a fancy school (located on the Singel, one of the main canal streets of Amsterdam), frequented by the children of well-to-do Dutch merchants who wanted their young to be worldly. Before opening his school Van den Enden ran a bookshop and art gallery, In de Kunst-Winkel, that was quite an attractive meeting place for intelligent youths yearning for unconventional ideas. With his energy and erudition, Van den Enden cut a charismatic figure and it is easy to imagine him as a genial and sly leader of young political and religious dissidents. (He was around fifty when Spinoza met him and seventy when he was hanged in France in the aftermath of a failed plot to overthrow Louis XIV. He spoke good French, but was not aristocratic enough to deserve the glory of the guillotine.)

Spinoza first joined the Van den Enden school because he needed to learn Latin, the *lingua franca* of philosophy and science that his otherwise broad education had not yet included. But he did not just learn Latin at the school. He learned of philosophy, medicine, physics, history, and politics, including those of free love that the libertine Van den Enden advocated. Spinoza must have approached this shop of forbidden pleasures with abandon and delight. Van den Enden's was a school for scandal if ever there was one, and it also seems to have given Spinoza his first taste of love in the person of his young Latin tutor, Clara Maria Van den Enden.

Acquaintance with Van den Enden produced a notable inflection in Spinoza's life at a time when other personal changes were taking place. In the few years that preceded his enrollment, at about age seventeen or eighteen, Spinoza had become an active businessman in his father's firm. Entering the world of business meant interrupting his formal studies, although he remained a part of synagogue life and appears to have joined a discussion group led by Rabbi ben Israel, the sort of intellectual gathering that would have been accessible only to advanced students of Judaism. Entering the world of commerce also meant encountering like-minded young business colleagues who were not Jewish. They included Jarig Jelles, a Mennonite in his thirties, Pieter Balling, a Catholic of unknown age, and Simon de Vries, a Quaker three years younger than Spinoza. The three men did not have Spinoza's intellectual caliber, but they shared a dissident streak, religiously and politically; an avidity to debate new ideas; and a youthful appetite for life. Juan de Prado, the only Jewish contemporary Spinoza befriended, was another young dissident repeatedly censored by the synagogue for his heretic comments and eventually expelled as well. The stage was set for a major influence of the new and secular on Spinoza's barely beginning adulthood.

The influence of the new has to be seen in the perspective of the old. The new ideas of Spinoza's questioning age sharply conflicted with the old ideas of the community in which he had been educated. Spinoza studied the Talmud and the Torah, and read the Kabbalah texts that came out of the Sephardic tradition and were especially popular among the Portuguese Jews of Amsterdam. The clash could hardly have been more dramatic. There were miracles in the old texts, but scientific explanations for those miracles could be formulated from the new facts. There was a blind trust in mystery and hidden meanings in old texts, but new evidence made it possible to explain the mysteries. The old superstitions could be exposed for what they were.

The clash might have been inevitable, but Spinoza's personal history made it all the more likely. Spinoza's mother died when he was six and she not yet thirty, and her loss was another shadow in this otherwise fortunate upbringing.[5] Not much is known about her, but her contribution to the development of young Spinoza is likely to have been considerable, and her death a deeply felt event. I do not imagine there was much childhood left after that, if ever such a childhood had been meant to be. Descriptions of the ten-year-old Spinoza helping with his father's business while frequenting school leave the impression of a premature adulthood. The boy was exposed to the real world of commerce and to the glories and frailties of human beings struggling to make a living in the teeming microcosm of Amsterdam. Miguel de Espinoza remarried three years after Spinoza's mother died, and the closeness to his father seems to have increased. The story goes that in spite of his active participation in the community's religious life, Miguel had little patience for hypocritic behavior, religious or not. He derided ceremonial piety and taught his son how to tell true from false when it came to human relations. Not surprisingly, young Spinoza despised superstition and artificiality. He was noticeably cocky, and his wit often embarrassed his teachers. Also, Miguel never made a secret of his skepticism on the matter of the immortal soul. Spinoza was certainly primed to see beyond the façade of piety and must have become alert to the large distance between the prescriptions of the religious texts and the daily practices of common mortals. Spinoza's questioning of the merits of rituals seems to have begun at home.

The Uriel da Costa Affair

Perhaps the beginning of Spinoza's rebellion can be traced to events that marked the last year in the life of Uriel da Costa, a relative of Spinoza's on his mother's side, and a central figure of the Jewish community in Amsterdam during Spinoza's boyhood.

The critical episode took place in 1640, according to some sources, or in 1647, according to others, meaning that Spinoza might have been as young as eight but not older than fifteen. So here are the antecedents.

Uriel da Costa had been born Gabriel da Costa in Porto, the Portuguese city from where Spinoza's mother came. His was also a family of rich Sephardic merchants that outwardly converted to Catholicism. Gabriel was raised as a Catholic and enjoyed a privileged life. He was an aristocratic young gentleman who grew up with two passions, horses and ideas, and whose intellectual bent led to a career at the University of Coimbra where he studied religion and became a professor. However, as the brooding da Costa deepened his knowledge of religion, he found increasing fault with Catholicism and gradually concluded that the ancestral Jewish faith of his family was truer and far more preferable. These conclusions should have been kept secret, but perhaps they were not. Da Costa and his mother, perhaps other relatives, went from being *conversos*—Jews converted to Christianity, to being *marranos*—Christians who practiced Judaism secretly. With or without justification, da Costa sensed that the long shadow of the Inquisition was upon him and he became convinced that he and his family were in danger. He persuaded them to move to Holland. All three brothers, his mother, and his wife, their servants and their caged birds, the elaborate furniture, the delicate china, and the rich linen that filled their manorial Porto residence and the summer house boarded a boat on the Douro River under the cover of the night.[6] And off they were, like so many before and after, bound up the Atlantic coast for a Dutch or German harbor and a new life.

I am telling this long preamble so that I can announce that
after establishing himself in Amsterdam, shedding his Portuguese
given name, Gabriel, and adopting the Hebrew variant, Uriel, da
Costa took to the fine analysis of Judaism and some more brooding. This time he found fault with Jewish practices and teachings
and became openly vocal about his findings: Religious practices
were superstitious; God could not possibly be humanlike; salvation should not be based on fear, and so on. All this and more he
not only said but wrote. The synagogue responded with the expected criticisms and admonishments. Over the ensuing decades,
da Costa was excommunicated, then reinstated, and then excommunicated again, finding refuge at some point in the Jewish community of Hamburg from which he eventually was expelled as
well. The da Costa affair had become a serious matter for the Jewish nation because its leaders feared that a blatant heresy such
as da Costa's would discredit the community or worse. The Dutch
authorities might consider reprisals against the entire group
based on their fear that antireligious Jewish feeling might propa-
gate to the Protestant population.

By 1640 (or 1647 at the latest) the da Costa saga came to a
head. The synagogue wanted a solution to this embarrassing

episode and so did da Costa who was then in his mid-fifties and noticeably consumed, physically and mentally, by this never-ending battle. A settlement was reached. Da Costa was to come to the synagogue and recant his heresy so that all could witness his repentance. Then he would be punished physically so that the serious nature of his crime would not be missed. Then he could regain his status in the Jewish nation.

In his book *Exemplar Vitae Humanae*, da Costa rebels against this prepotence and leaves no doubt that his acceptance of the settlement does not mean his ideas have changed at all. He makes clear, however, that the continued humiliation and sheer physical exhaustion left him no other choice.

The punishment day was thoroughly publicized and eagerly anticipated, great theater and great circus rolled into one. The synagogue was packed with men, women, and children sitting and standing with hardly any space to move, all waiting for the unusual entertainment to begin. The air was thick with excited breath and the silence was broken only by the grating of shoes against the sand grains that covered the wooden floors.

At the appropriate moment da Costa was asked to climb to the central stage and invited to read a statement prepared by the leaders of the congregation. Using their words, he confessed his numerous transgressions, the nonobservance of the Sabbath, the nonobservance of the Law, the attempt to prevent others from joining the Jewish faith, all of which warranted a thousand deaths, but were to be forgiven because he promised, in reparation, not to engage ever again in such odious inequities and perversities.

Once the reading was over, he was asked to step down from the stage and a rabbi whispered in his ear that he should now move to a certain corner of the synagogue. He did. At the corner, the *chamach* asked him to undress down to the waist, take off his shoes, and tie a red handkerchief around his head. He was then made to lean against a column and his hands were tied to it with a rope. Now the silence was sepulchral. The *hazan* approached, leather whip in hand, and began to apply thirty-nine slashings to da Costa's bare back. As the punishment proceeded, perhaps to pace the slashings, the congregation began singing a psalm. Da Costa counted the slashings and credited his torturers with a scrupulous observance of the Law, which specified that the number of strokes should never exceed forty.

The punishment over, da Costa was allowed to sit on the floor and put his clothes back on. Then a rabbi announced his reinstatement for all to hear. The excommunication was lifted and the door to the synagogue was now as open to da Costa as the door to heaven would be one day. We are not told if the news was greeted with silence or applause. Silence is my guess.

But the ritual was not complete yet. Da Costa was now asked to come to the main door and lie down on the floor across the threshold. The *chamach* helped him to the ground and held his head in his hands with solicitude and gentleness. Then, one by one, men, women, and children left the temple, and each person had to step over him on the way out. No one actually stepped *on* him, he assures us in his memoir, just *over*.

Now the synagogue was empty. The *chamach* and a few others congratulated him warmly on a punishment well taken, and on the arrival of a new day in his life. They helped him up, and dusted the sand that had dropped from so many shoes onto his tattered clothes. Uriel da Costa was once again a member in good standing in the New Jerusalem.

It is not clear exactly how many days this accommodation lasted. Da Costa was taken home and proceeded to finish his manuscript of *Exemplar Vitae Humanae*. The last ten pages deal with this episode and his impotent rebellion against it. After finishing the manuscript, da Costa shot himself. The first bullet

missed its target, but the second killed him. He had had the last
word in more ways than one.

Nowhere in his books or in the correspondence that survived him
did Spinoza ever mention Uriel da Costa by name. And yet Spinoza knew all about da Costa. It is true that there were other ex-
communications, recantations, and public punishments during
this same period. In 1639, a man named Abraham Mendes was
subject to a similar punishment—recanting, lashings, and walk-
over by the community—suggesting that the synagogue did not
hesitate to impose discipline among its ranks.7 But the da Costa
affair was easily the most salient of its kind. He was not a simple
heretic, but rather a published heretic, and he persisted in his
wrong ways for decades, which explains the atmosphere of scandal. Spinoza, whether age eight or fifteen, was in the audience
with his father and siblings. Moreover, for years on end, the case
was talked about as a reference, and one senses its contours in
some of Spinoza's writings about organized religion. Finally, and
perhaps most importantly, Uriel da Costa's general position in relation to organized religion became Spinoza's position as well.8
Da Costa was not a deep thinker like Spinoza. He was a troubled
man who could not stop himself from suffering with every perceived iniquity and responding with indignation. He gave voice to
a perception of hypocrisy that was shared by many others at the
time, and his real originality was martyrdom. It is possible that
Spinoza's silence on this affair reflected his decision to deny an influence of da Costa's ideas given that such ideas were in the air
anyway and da Costa never treated them with the depth of analysis that Spinoza eventually did. Or maybe that Spinoza simply suffered from anxiety of influence and would not acknowledge a
debt, if there was one, consciously or unconsciously. (The same

could be said, incidentally, of his relation to Van den Enden. Spinoza never cited him by name.) Be that as it may, it is reasonable to believe that the da Costa affair had an immense impact on Spinoza, more so because of its drama than the analyses expressed in *Exemplar Vitae Humanae*. Memories of the episode possibly stiffened Spinoza's back for his own combat to come, and must have guided his decision not to be present for his own excommunication. Spinoza's *cherem* was read on the same stage as da Costa's recantation but *in absentia*.

Jewish Persecution and the Marrano Tradition

Despite its outward prosperity, the Jewish nation of Amsterdam was not quite secure. There was a persisting fear that any false move by a Jew could be misinterpreted by the Calvinist authorities and result in criticism or punishment of the community. The Jews were used to persecution, and the gentlemen's agreement under which they were living in Amsterdam required a fine toeing of the line. There had to be a public display of faith in God, but no public defense of Judaism and no attempt to bring a local citizen into the Jewish faith. There could be no marriages to the local citizens. Most of all, there had to be discretion.

The Jews were useful guests, not compatriots. Their good behavior could be rewarded with civil liberties, but the risk of losing such liberties hovered over them. Uriel da Costa's punishment was designed to remind the community of that risk. Spinoza's own generation of Jews probably regarded itself as Dutch, not as exiles, and Spinoza does assume a Dutch identity with the passage of time. But the grounds for that identity were recent and not especially firm.

The architecture of the new Portuguese synagogue in Amsterdam told it all. The remarkable structure, which opened its doors in 1675, is not a single building but a walled compound designed to contain a sanctuary, school, and grounds for adults to meet and children to play protected from the society around it.

Community leaders had real concerns over possible violation of rules set by the Dutch hosts. First, the leaders realized that while the welcome they received was predicated on Dutch business interests, the firmness of the welcome depended on the notably tolerant and generous attitude of a segment of the Dutch authorities. The size of that segment varied with the vagaries of politics, and the benign influences shrunk or enlarged accordingly. While DeWitt was grand pensionary, for example, the Dutch provinces functioned as the most advanced democratic republic of the age. The more conservative and bigoted influences (the Orangists) were held in check. But the opposite occurred after DeWitt's assassination in 1672 and the democratic dream was suspended.

Second, in spite of a considerable cohesion, there were tensions inside the Jewish community. For example, there were conflicts related to religious practices, which is not surprising since most, perhaps all, of the members of the *nação* had practiced in secret in Portugal without the help of a synagogue. And there were conflicts related to a host of social issues, again unsurprising and inevitable in a traditionally segregated group. The leaders of the *nação* did everything possible to prevent these conflicts from becoming visible to the Dutch. The image of God-loving and hardworking people they wished to project could not be shattered. It was embarrassing enough to contend with the social consequences of the Sephardin's sexual appetite, which was reputed to be insatiable. Or to manage the presence of the very different group of Jewish immigrants that came from the north and east of Europe and were mostly poor and uneducated. Spinoza grew up as an attentive witness to human conflict, intrapersonal, social, religious, and political. When he wrote of human beings and their weaknesses, alone or within the religious and political institutions they created, he knew whereof he spoke.

Spinoza was keenly aware of the history of the Sephardic Jews prior to their arrival in the low countries and thoroughly familiar with the religious and political dimensions of the Jewish problem—he commented on it in the *Tractatus*. The choice and shaping of the themes of his philosophy could not escape the weight of this history, and the *marranos* are an important part of this history.

The *marrano* tradition consisted of the secret practice of Jewish rituals by Jews forced to convert to Christianity. This tradition began in 1492 in Spain, decades before the time the Jews were expelled, but became especially intense in Portugal after 1500. It still was going strong a century later, at the time the elite of the community was making its way to the low countries.[9]

After 1492 the Spanish Sephardic Jews fled to Portugal in large numbers. By some counts more than 100,000 crossed the border, attracted by the peaceful way in which Portugal had treated the Jews until then. The Portuguese Jewish community, however, was modest in size and the sudden increase in its numbers introduced a host of social problems. At stake was the issue of how to weave the new population into the fabric of Portuguese society. The wealth and status of a significant part of the new group—most were merchants, financiers, professionals, and skilled artisans—sharply distinguished them from the small Portuguese bourgeoisie of the time, as much as it distinguished them from the ordinary people and from the aristocracy. They did not quite fit. Amid much disquiet, King John II and his successor King Manuel I tried to deal with the problem with widely different strategies. In 1492, when the issue first arose, John II taxed the new arrivals to the hilt. All that eight cruzados got them, per head, was a sojourn in Portugal for just as many months. Beyond that period, the newcomers had to pay a huge and undisclosed tax to the crown for a permanent stay to be granted. Failing that, the fugitives were accorded no civil rights and no citizenship. In effect, they belonged to the king and existed at his pleasure. John II's successor, Manuel I, took a different tack. Portugal was engaged in a colossal colonial enterprise, the building of an overseas empire entirely out of proportion with the limited size of its land and its population. Manuel I recognized the potential value of the Jews in the running of this extraordinary effort. Accordingly, he restored their civil rights. The downside of this good measure, however, was the exorbitant price: The Jews were forced to convert to Christianity. They had to be baptized or leave the country.[10]

In short order, many Jews who were first expelled and then exploited were now baptized. What really happened afterward is difficult to describe in terms of the numbers, but goes roughly like this. A significant part of the Sephardic population was fully assimilated to Christianity, Portuguese style, with varying degree of distress. They became *conversos* or *cristãos-novos* (New Christians). Their descendants can be found today, after many generations, as Catholics, Protestants, or without religious ties. They blend into the life of this old country, their Jewish origin obscured by the passage of five centuries. Another part of the Sephardic population became *marrano*. Outwardly the *marranos* behaved as Christians, but they struggled to remain observant Jews behind closed doors and keep their traditions alive. It is unlikely that most New Christians were secret practitioners, but no one knows how many there were or for how long they practiced. Incidentally, the term *marrano*, from the Spanish *marrar*, is both a plain insult (it stands for swinish) and an intellectual slight (it also means incomplete or failure).

The fate of the *marranos* is quite varied. Some of them per-
ished at the hands of the Inquisition, which eventually was established in Portugal (1536)[11] and turned its attention away from
Protestant heretics—there were not many to persecute in Portugal—to the *marranos,* a far more lucrative endeavor for the
church and for the state.[12] Another group of the *marranos* gave up
their brave resolve to maintain a risky and waning historical tradition, and they too joined the ranks of former Portuguese Jews.
The smallest segment of the *marranos* eventually left Portugal, the
segment whose extensive wealth and international contacts allowed them to emigrate.

The *marranos* changed their names frequently, not just
for symbolic reasons—such as when Gabriel renamed himself
Uriel—but for protection. Aliases confused the Inquisition's
spies and delayed the casting of suspicions over family members
still in Portugal. The fact that not just activities but also ideas had
to be concealed was fresh in the minds of the adults around
whom Spinoza was growing up. A stoic attitude is another legacy
of *marrano* life. Life in general and the faith in particular had been
maintained in trying circumstances for many decades without the
help of a religious institution—the synagogues were closed, of
course—and with a courageous self-effacement. Eventually, when
Spinoza had to conceal his own ideas, the reasons were not altogether different, and the ancestral experience was useful. The
tradition of artful disguise came naturally, and so did the stoic
streak, which is a defining trait of Spinoza's human praxis and
whose origins need not be sought only in Greek philosophy. Most
importantly, however, the recent history of Sephardic Jews forced
Spinoza to confront the strange combination of religious and political decisions that had maintained the coherence of his people
through the centuries. I believe the confrontation led Spinoza to
take a position on that history. The result was the formulation of

an ambitious view of human nature that might transcend the problems faced by the Jewish people and be applicable to humanity at large.

Would Spinoza have been the same without the giddy sense of liberation that the *marranos* experienced in Amsterdam? Not quite, I imagine. Would Spinoza have been Spinoza had his parents remained in Portugal? Can one imagine Bento growing up in Porto, Vidigueira, or Belmonte? Not a chance, of course, for one and a thousand reasons. It is true that the conflict inherent in the *marrano* mind moves it away from irreconcilable religious forces and toward the natural and the secular.[13] But whatever the intensity of the *marrano* conflict, a spark was needed to light the fire of creativity, and that spark was freedom. This may sound paradoxical given the way Holland treated Spinoza's work after his death, but it is not. Dutch freedom was not quite ample enough to accommodate, let alone welcome, Spinoza's work once it was ready and published. But it was ample enough to allow him access to new and relevant reading material of his own time; ample enough for him to debate his new ideas with others of varied religious and social backgrounds; and ample enough, albeit just barely, for him to be an independent entity dedicated to the single activity of rethinking human nature. None of this would have been possible in Portugal or, for that matter, anywhere in the world in the seventeenth century. The unique environment of the Dutch Golden Age was needed to turn the pent-up conflicts of a penalized people into the creative exuberance of a gifted human being.

Excommunication

Spinoza was born into a community of exiles and by age twenty-four had become exiled from that very community. He was on his way to an even greater physical and social isolation that was transcended only by the universal character of his work. The events surrounding the final chapter of his relation to the synagogue are almost as dramatic as Uriel da Costa's. The rabbis knew of Spinoza's ideas and were aware that he was developing arguments against many aspects of the Law. Until the death of his father, however, except for the debates he engaged in with individual rabbis, Spinoza seemed to have given little public airing to his ideas and did not commit them to writing. He continued to come to the synagogue, and since his father died the twenty-two-year-old Spinoza had assumed responsibility for the firm. The break took place at that point. He became more vocal and was no longer afraid of the embarrassment his opinions caused; he made close friendships outside the fold; and he began taking secular matters involving members of the Jewish community into the secular Dutch world. Spinoza ignored the hard rule that all social issues pertaining to Jews—disputes over business, and property, and the like—should be handled within the secular arm of the *nação,* and not in the Dutch courts.

The elders of the synagogue used every means at their disposal to persuade him to think and behave differently. They promised an annuity of one thousand florins, and one can only imagine the barely polite contempt with which Spinoza declined the offer. Later they issued a "lower" excommunication, which separated Spinoza from the community for thirty days. Even later they may have ordered the murder attempt, which Spinoza survived. The maneuvering simply strengthened Spinoza's resolve.

On July 27, 1656, the synagogue finally issued the "high" *cherem.* A few words about the phenomenon are in order. It is important to note that while *cherem* is always translated as excommunication, a more accurate translation of the word is banishment or exclusion. The punishments were not handed over by ecclesiastical authorities, but rather by community elders, the *senhores* or "councilmen," although the rabbis were consulted and the consequences were not only religious. The recipient of a *cherem* was excluded from the community in physical and social terms. On the other hand, one must note how relatively mild the *cherem* was in comparison with the Catholic equivalent: the *auto-da-fé*. Even the thirty-nine lashings of poor Uriel da Costa pale when compared to the torture chamber and burning stake, which were the fate of unrepentant heretics at the hands of the Inquisition, whether or not they had anything to repent. Evil has many degrees, after all.

By the standards of Amsterdam's Jewish community, Spinoza's *cherem* was considered cruel and unusual, violent and destructive. There also is little doubt that the community was embarrassed by this punishment. When Johannes Colerus, Spinoza's main contemporary biographer, first tried to obtain the text of the *cherem,* the elders stonewalled.

The records of the community—*O Livro dos Acordos da Nação*—show that there were fifteen "severe" *cherems* handed down from Spinoza's birth to the time of his own *cherem.* None of the others were as violent in language or as thorough in its condemnation. Curiously, the anathema that is part of Spinoza's *cherem* seems to have been written decades earlier by the elders of the Venice Sephardic community. This anathema was imported by the Amsterdam elders long before 1656 and included in a book of punishment recipes to be used as needed in cases of indiscipline. Rabbi Mortera, former mentor of Spinoza and close friend of his father, had selected this particular menu item for Spinoza. The text is worth reproducing, in a translation of the Portuguese original provided by the Spinoza scholar Frederick Pollock in 1880.

> The chiefs of the council do you to wit, that having long known the evil opinions and works of Baruch de Espinoza, they have endeavoured by diverse ways and promises to

withdraw him from his evil ways, and they are unable to find a remedy, but on the contrary have had every day more knowledge of the abominable heresies practised and taught by him, and of other enormities committed by him, and have of this many trustworthy witnesses, who have deposed and borne witness in the presence of the said Espinoza, and by whom he stood convicted; all which having been examined in the presence of the elders it has been determined with their assent that the said Espinoza should be excommunicated and cut off from the nation of Israel; and now he is hereby excommunicated with the following anathema:

With the judgment of the angels and of the saints we excommunicate, cut off, curse, and anathematize Baruch de Espinoza, with the consent of the elders and of all this holy congregation, in the presence of the holy books: by the 613 precepts which are written therein, with the anathema wherewith Joshua cursed Jericho, with the curse which Elisha laid upon the children, and with all the curses which are written in the law. Cursed be he by day and cursed be he by night. Cursed be he in sleeping and cursed be he in waking, cursed in going out and cursed in coming in. The Lord shall not pardon him, the wrath and fury of the Lord shall henceforth be kindled against this man, and shall lay upon him all the curses which are written in the book of the law. The Lord shall destroy his name under the sun, and cut him off for his undoing from all the tribes of Israel, with all the curses of the firmament which are written in the book of the law. But you that cleave unto the Lord your God, live all of you this day.

And we warn you, that none may speak with him by work of mouth nor by writing, nor show any favour to him, nor be under one roof with him, nor come within four cubits of him, nor read any paper composed or written by him.[14]

Thus Spinoza was severed from the community. Jewish acquaintances and family were forbidden to see him and had to stay away. He was free as a bird and almost as dispossessed. He called himself Benedictus now.

It should be noted that even in this phase of open scandal, there was no hint that Spinoza tried to exploit the embarrassment of his judges to win a public victory with his words. He probably could have exposed the prepotence of the synagogue, had he wished, and responded to the *cherem* with a barrage of rhetorically devastating arguments, but he did not.[15]

Spinoza's restraint was an early sign of the wisdom that led him, years later, to insist that his texts should be available only in Latin, so that only those who were sufficiently learned could read them and contend with the potentially troubling ideas they conveyed. I believe Spinoza was genuinely concerned with the impact his ideas might have on those who had nothing but faith to maintain the balance of their lives.

On the high summer day of July 27, 1656, probably at the house of a Dutch friend not far from the synagogue, Spinoza is presumed to have greeted the news of the *cherem* with these words: "This compels me to nothing that I should not otherwise have done." Simple, dignified, to the point.

The Legacy

Spinoza's legacy is a sad and complicated matter. One might say that given the historical context and the uncompromising positions he took, the vehemence of the attacks and the efficacy of the prohibition of his work should have been anticipated. To a certain degree Spinoza did, as his precautions seem to indicate. Still, the reaction turned out to be stronger than anyone might have expected.

Spinoza left no will, but he had given Rieuwertz, his friend

and publisher in Amsterdam, detailed instructions for the disposition of his manuscripts. Rieuwertz was nothing if not loyal, and
he also was courageous and very smart. Spinoza died in late February of 1677, but by the end of the same year a book entitled
Opera Posthuma was printed, with *The Ethics* at the heart of the
volume. Dutch and French translations began to appear in 1678.
Rieuwertz, and the band of Spinoza's friends who helped him,
had to contend with the most violent outrage yet against Spinoza's ideas. The condemnation of the Jews, the Vatican, and the
Calvinists was expected, of course, but the reaction went further.
Dutch authorities were the first to ban the book, and then other
European countries did too. In a number of places, namely in
Holland, the ban was firmly enforced. Authorities inspected
bookshops and confiscated any volumes they found. Publishing
or selling the book was an offense and it remained an offense for
as long as there was any curiosity about it. Rieuwertz eluded the
authorities in a masterful way, consistently denying any knowledge of the originals and any responsibility for the printings. He
managed to distribute a number of books illegally, in Holland
and abroad, exactly how many is not clear.

Spinoza's words were thus secure in many private libraries of
Europe, in clear defiance of the churches and the authorities. In
France, in particular, he was widely read. There is no doubt that
the more accessible aspects of the work—the part that dealt with
organized religion and its relation to the state—were being absorbed and in many quarters admired. Nonetheless, the churches
and authorities largely won their battle because Spinoza's ideas
could hardly be quoted in print in a positive light. The injunction
was implicit rather than openly legislated, but it produced even
better results that way. Few philosophers or scientists dared side
with Spinoza because that would be courting disaster. Supporting

any claim by openly citing Spinoza's arguments or tracing an idea to his texts would undermine the chances of having the claim heard. Spinoza was anathema. This applied throughout Europe for the best part of the hundred years that followed Spinoza's death. On the contrary, negative references were welcome and abundant. In some places, as was the case of Portugal, mentions of Spinoza carried an obligate pejorative qualifier, such as "shameless," "pestilent," "impious," or "stupid"![16] On occasion the critical views were smoke screens and managed to disseminate Spinoza's ideas in covert fashion. The most notable example of that mock dubiousness was Pierre Bayle's article about Spinoza in the *Dictionnaire Philosophique et Critique*. Maria Luisa Ribeiro Ferreira contends that Bayle was at most ambivalent and perhaps purposefully ambiguous in his text; he actually managed to call attention to Spinoza's views while appearing to dismiss his ideas.[17] Notably, the Spinoza entry is the largest in the entire dictionary.

On occasion, however, clever dubiousness and ambivalence were not allowed to stand and the secret admirers were urged to expunge their writings of impious Spinozism. Or else. A prominent example regards *L'Esprit des Lois*, Montesquieu's major contribution to the Enlightenment (1748). Montesquieu's views on ethics, God, organized religion, and politics are through and through Spinozian and were, unsurprisingly, denounced as such. Montesquieu seems not to have anticipated the destructiveness of the attacks. Not long after the publication Montesquieu was forced to deny the Spinozism of his views and make a public declaration of his faith in a Christian creator God. How could such a believer have anything to do with Spinoza? As Jonathan Israel recounts the episode, reservations about Montesquieu persisted and the Vatican remained unconvinced. *Caute!*

As the record was purified of Spinoza references, his ideas became progressively more anonymous for future generations. Spinoza's influence was *unacknowledged.* Spinoza was pilloried and plundered. In life his identity was known but his ideas were *sub rosa;* after death the ideas float freely but the author's identity was obvious only to contemporaries and carefully concealed from the future.

This state of affairs is finally changing. Recently, it has become clear that Spinoza's work is a decisive engine behind the development of the Enlightenment, and that his ideas helped shape the central intellectual debate of eighteenth-century Europe, although the history of the period would hardly lead anyone to believe that. Jonathan Israel makes this case convincingly and reveals important facts behind the silence that has led so many to believe that Spinoza's influence died with him.[18] Israel offers evidence against the widespread impression that John Locke's work dominated the debate from the very early phases of the Enlightenment. For example, one of the centerpiece publications of the Enlightenment, Diderot and d'Alembert's *Encyclopédie* dedicated five times more space to Spinoza than Locke, although it lavishes more praise on Locke, perhaps, as Israel suggests, with "a diversionary purpose." Israel also points out that in Johann Heinrich Zedler's 1750 *Grosses Universal Lexicon*—the largest eighteenth-century encyclopedia—the entries for "Spinoza" and "Spinozism" are each larger than the modest entry for Locke. Locke's star does rise, but later.[19]

Sadly, few philosophers of sound mind, young or old, ever paid public homage to Spinoza, let alone assumed the role of disciple or continuator. Not even Leibniz did so, although he read all of Spinoza's writings before they were published and probably was the best-qualified mind to appreciate Spinoza at the time. He ran for cover, like most everyone else, and adopted a measured critical position instead. The official lights of the Enlightenment did likewise. Privately they were illuminated by Spinoza; publicly they decried him. Voltaire's little poem on Spinoza exemplifies the mandatory public criticism and personal ambivalence toward the philosopher.[20] In my translation, the poem reads as follows:

> And then, a little Jew, with a long nose and a pale complexion,
> Poor but satisfied, pensive and reserved,
> A subtle but hollow spirit, less read than celebrated,
> Hidden under the mantle of Descartes, his mentor,
> Walking with measured steps, comes close to the great
>   being:
> Excuse me, he says, addressing him in a whisper,
> But I think, just between us, that you do not exist at all.
>   —Voltaire

Beyond the Enlightenment

After the Enlightenment the Spinoza influence became more open. Citing Spinoza was no longer offensive. There is a growing secular world that turns Spinoza into its prophet, "usually little read, poorly read, or not read at all," as Gabriel Albiac put it quite accurately.[21] But some read him and live by his lights. Philosophers such as Friedrich Heinrich Jacobi, Friedrich von Hardenberg Novalis, and Gotthold Lessing introduced the thinker to a different audience and a different century. Goethe adopted Spinoza and became his champion, leaving no doubt as to Spinoza's influence on his person and his work. "This man, who wrought wonderfully on me and who was destined to affect so deeply my entire mode of thinking, was Spinoza. After having looked round the world in vain for means of developing my nature, I met with the *Ethics* of that philosopher. Of what I read *in* the work, and what I read *into* it, I can give no account; but I found in it a sedative for my passions, and it seemed to unveil a clear broad view over the material and moral world. But what essentially riveted me to him was the boundless disinterestedness which shone forth in every sentence. That wonderful sentiment, 'He who loves God must not expect God to love him in return.'"[22]

The English poets become equally vocal champions. Samuel Taylor Coleridge absorbed Spinoza, and so did William Wordsworth, drunk with nature of his own accord and drunk with Spinoza's own inebriation with the divine in nature. And so did Percy Shelley, Alfred Lord Tennyson, and George Eliot. Spinoza might have reentered philosophy earlier had Kant not declined to read him, and had David Hume been more patient. Eventually Georg Hegel proclaimed, "To be a philosopher you must first be a Spinozist: if you have no Spinozism, you have no philosophy."[23]

Spinoza's influence on the fields of contemporary science most naturally linked to his ideas, biology and cognitive science, appears all but absent. But this was clearly not the case in the nineteenth century when Wilhelm Wundt and Herman von Helmholtz, two of the founders of the mind and brain sciences, were avid followers of Spinoza. In reading through the list of international scientists who joined in 1876 to erect the statue of Spinoza that now sits in The Hague, I found both Wundt and Helmholtz, as well as Claude Bernard.[24] Could it be that Spinoza inspired Bernard's preoccupation with the notion of a balanced life state?

In 1880 the physiologist Johannes Müller noted "the striking resemblance between the scientific results attained by Spinoza two centuries ago, and those reached in our own days by workers, who, like Wundt and [Ernst] Haeckel in Germany, [Hippolyte] Taine in France, and [Alfred] Wallace and Darwin in England, have come to psychological questions through physiology."[25] My suggestion that Spinoza was a forerunner of modern biological thinking appeared obvious to Müller, and to Frederick Pollock, who said, at about the same time, that Spinoza "tends more and more to become the philosopher of men of science."[26]

The acknowledgment seems to dry up again in the twentieth century. For example, Spinoza appears to have had an important influence on Freud. Freud's system requires the self-preservation apparatus Spinoza proposed in his *conatus,* and makes abundant use of the idea that self-preserving actions are engaged non-consciously. Yet Freud never cited the philosopher. When questioned on this issue, Freud took great pains to explain the omission. In a letter to Lothar Bickel in 1931 Freud wrote: "I confess without hesitation my dependence regarding the teachings of Spinoza. If I never cared to cite his name directly, it is because I never drew the tenets of my thinking from the study of that author but rather from the atmosphere he created."[27] In 1932 Freud closed the door once and for all on any acknowledgment. In another letter, this time to Siegfried Hessing: "I have had, for my entire life, an extraordinary esteem for the person and for the thinking of that great philosopher. But I do not believe that attitude gives me the right to say anything publically about him, for the good reason that I would have nothing to say that has not been said by others."[28] In fairness to Freud we should recall that Spinoza acknowledged neither Van den Enden nor da Costa. Perhaps if he had been questioned on that omission his answer might have resembled Freud's.

Three decades later the noted French psychoanalyst Jacques Lacan dealt with the Spinoza influence in a somewhat different manner. In his 1964 inaugural lecture at the École Normal Supérieure, suggestively titled "The Excommunication," Lacan recounted how the International Psychoanalytical Association attempted to prevent him from training psychoanalysts and expelled him from its ranks. He compared this decision to a grand excommunication and reminded his audience that this was precisely the punishment Spinoza received on July 27, 1656.[29]

There is one important exception to all this denial of the father. Albert Einstein, the emblematic scientist of the twentieth century, did not hesitate to say that Spinoza had a profound influence on him. Einstein felt quite comfortable with Spinoza's views on the universe in general and God in particular.[30]

The Hague, 1677

Spinoza died midway through his forty-fourth year. He had suffered from a respiratory ailment for years. His chronic cough is well documented, and he also smoked regularly. The pipe was his visible concession to the world of sensuous pleasures. Besides, he may have believed the claim that tobacco conferred some protection against the plagues that ravaged Europe during his life. Spinoza survived several epidemics of plague that killed many around him. Maybe the smoking helped. In the months before his death his condition worsened, although he never stopped working and receiving visitors. His death was unexpected. He expired during the afternoon of Sunday, February 21, but in the morning of this last day he had come down to lunch with the Van der Spijk family, as was his custom. The family was away at church that Sunday afternoon, but Spinoza was attended by his Amsterdam physician, Ludowick Meyer, when he passed away.

Spinoza's demise is usually attributed to tuberculosis, but there is no evidence that he was consumptive. In all probability, his illness was far less commonplace. He may have succumbed to a professional hazard, silicosis, as suggested by Margaret Gullan-Whur.[31] Silicosis, a disease that was not yet recognized, is caused by inhaling dust from the grinding of glass, and that is precisely the activity Spinoza seems to have pursued most days of his adult life. Unprotected by the face mask he would have worn today, and without the help of either consumption or the plague, Spinoza may have lined the inside of his lungs with shiny glass powder until breath was no longer possible.

By then the confidence with which he arrived in The Hague had become even stronger and taken the form of unshakable convictions. But the dream of recognition and influence, if he ever entertained one seriously, had vanished entirely. In its place were tranquillity and acceptance.

The Library

Back inside the Rijnsburg house I look again at Spinoza's books. There is Machiavelli, De Grotius, and the Thomases—More and Hobbes—the marriage of the art of politics with the art of jurisprudence. There is Calvin, several Bibles, a book on the Kaballah, and many dictionaries and grammars, the basics of home reference. There are books on anatomy—Dr. Tulp's, he of Rembrandt fame, and Dr. Kerckring's. Theodor Kerckring was Spinoza's contemporary, colleague and rival. He, too, was a student at Van den Enden's school. He, too, was besotted with Clara Van den Enden, but he walked away with Clara and with her to the altar. It was a nice touch that Spinoza kept his two books. I can imagine that Spinoza forgave them both and forgot all about the necklace that Theodor gave Clara, when our empty-handed young prince could do nothing but cast his mournful eyes on the scintillating Clara.

The contemporary literature wing is sparse—Spain's Cervantes and Gongora are here but not Camões, Portugal's national poet. Is it conceivable that Spinoza would not have had Camões' *The Lusiad* nearby? Perhaps the books were pinched or perhaps he did not wish to be reminded of Portugal. Or perhaps he was not sensitive to modern verse. There are not many references to poetry, music, and painting in Spinoza, although Spinoza did acknowledge music, theater, the arts, and even sports as conducive to the happiness of the individual. Shakespeare and Christopher Marlowe are not here either, but then Spinoza did not read English and perhaps they were not translated. In this bookcase, even philosophy pales beside mathematics, physics, and astronomy, and only Descartes is properly represented.

It is a bit risky to judge a man's reading habits by the size and contents of his library, but somehow this bookcase rings true. Perhaps these were all the books he needed in his later years. The library is of a piece with the rest of his effects. It makes minimalism sound excessive. And then I go again through the book of visitors, find Einstein's entry, and try to imagine the scene of his visit to this room on November 2, 1920.

Spinoza in My Mind

Meeting Spinoza in my imagination was one reason for writing this book, but the meeting was a long time coming. My mind used to go blank whenever I thought of what Spinoza might have looked like, live and in action. This is not surprising. For one thing, the descriptions of his life are as discontinuous as his home address, and the reasonably contemporaneous biographies are not as rich in detail as one would wish. For another, Spinoza's style is hermetic. Some paragraphs of *The Ethics* and of the *Tractatus* can be devastatingly humorous. That is a clue, for certain. It also is true that Spinoza is never less than respectful of fellow humans, even those whose ideas he scorns. That is another clue, no doubt. Yet such clues are not enough to suggest a whole person. The man behind the writing is well sealed from the reader, either because of the limitations of his Latin or because Spinoza deliberately wanted to purge his texts of personal feeling and rhetoric. Stuart Hampshire inclines to the latter and I believe he is correct.[32] Gradually, however, from the quiet simmering of hints and

reflections, a flesh and blood portrait began to emerge in my imagination. Now I have no difficulty seeing Spinoza at different ages, in different places, in different situations.

In my story he begins as the impossible child, inquisitive, opinionated, a mind aged beyond its years. As an adolescent he is insufferably quick-witted and arrogant. He is at his worst around age twenty, when he is both a no-nonsense man of business and an aspiring philosopher; he sports the manner of an Iberian aristocrat but also is busy establishing his Dutch identity. This period of conflict is over by his mid-twenties. Suddenly he is no longer a Jew or a businessman; he has neither family nor home; but he is not defeated. He dominates small gatherings with the sharpness of his intellect and his enthusiasm. The legend of Spinoza the sage is born then. He also finds new occupations: the manufacture of lenses, which becomes his livelihood and deepens his study of optics; and drawing, a quiet pastime at which he seems to have excelled but for which we have no record.

By age thirty yet another transformation is under way. Spinoza begins to measure his steps. He curbs his wit. He is kinder than before to those around him and patient with the dimwits. The mature Spinoza is firm in his beliefs but less dogmatic, and even as he seems to become more tolerant of others he withdraws and seeks ever-quieter surroundings. This Spinoza of my imagination communicates stability to those around him. He is revered by many.

Do I *like* the Spinoza I finally met? The answer is not so simple. I admire him, for certain. I like him immensely, at times. But I wish I would be as clear about the ways of his mind as I am about the form of his behavior—something in him never yields to scrutiny and the strangeness about him never abates. I am clear enough, nonetheless, to marvel at the bravery with which he formulated his ideas at the time he did and adapted his life to the inevitable consequences. In his own terms he succeeded.[33]

CHAPTER 7

Who's There?

The Contented Life

Before Spinoza became clear in my mind, I used to ask myself a troubling question: Was Spinoza really content during his years in Voorburg and The Hague or was he posing for sainthood? Was he carefully constructing a persona of benignity and earthly denial that lent authority to his words and made the task of his critics all the more challenging? The Spinoza of my imagination answers this question easily. Spinoza *was* content. His frugality was not a ploy. He was not acting an example of sacrifice for posterity. His life and his philosophy probably fused around the ripe old age of thirty-three.

Assuming that Spinoza was content, and considering that his life lacked the trappings we usually associate with happiness— his poor health, no wealth, and lack of intimate human relationships would have prevented Aristotle from calling his life a success—it is sensible to ask how Spinoza achieved fulfillment. What was his secret? I am not moved by mere curiosity but rather by the opportunity of posing yet another question: How relevant to the achievement of a contented life is the knowledge of emotions, feelings, and mind-body biology that we have been discussing in this book? There is no doubt that emotions and feelings themselves are part and parcel of what we are, personally and socially. The question is: Does *knowing* how emotions and feelings work matter at all to how we live? I suggested earlier that such knowledge makes a difference to the governance of social life, but here I wonder if it may be just as pertinent to the inner circle of personal life governance.

Connecting this question to Spinoza is especially sensible when we note that the conception of human nature emerging under the influence of modern biology has some degree of overlap with Spinoza's own conception. We should definitely consider Spinoza's approach to contentment.

Spinoza's best-known recommendation for achieving a life well lived came in the form of a system for ethical behavior and a prescription for a democratic state. But Spinoza did not think that following ethical rules and the laws of a democratic state would be sufficient for the individual to achieve the highest form of contentment, the sustained joy that he equates with human salvation. My impression is that most human beings today probably would not think so either. Many people appear to require something more out of life beyond moral and law-abiding conduct; beyond the satisfaction of love, family, friendships, and good health; beyond the rewards that come from doing well whatever job one chooses (personal satisfaction, the approbation of others, honor, monetary compensation); beyond the pursuit of one's pleasures and the accumulation of possessions; and beyond an identification with country and humanity. Many human beings require something that involves, at the very least, some clarity about the meaning of one's life. Whether we articulate this need clearly or confusedly, it amounts to a yearning to know where we come from and where we are going, mostly the latter perhaps. What purpose greater than our immediate existence could life possibly have? And along with the yearning, there comes a response, in sharp focus or soft, and some purpose is either gleaned or desired.

Not every human being has such needs. The needs and wants of human beings vary to a considerable degree with their personalities, their inquisitiveness, their sociocultural circumstances, and even the time of their lives. Youth often leaves little time to consider the shortcomings of the human condition. Good fortune is another effective screen. Many would be puzzled by the mention of any extra requirement beyond youth, health, and good fortune. Why fret and fuss? Still, for those who recognize such needs, it is legitimate to inquire why they should long for something that may not come naturally or not come at all. Why are the extra knowledge and clarity desirable?

One might answer by stating that the yearning is a deep trait of the human mind. It is rooted in human brain design and the genetic pool that begets it, no less so than the deep traits that drive us with great curiosity toward a systematic exploration of our own being and of the world around it, the same traits that impel us to construct explanations for the objects and situations in that world. The evolutionary origin of the yearning is entirely plausible, but the account does require another factor so that one may understand why the human makeup ever incorporated the trait. I believe there was such a factor at work in early humans, much as it still is at work now. Its consistency has to do with the powerful biological mechanism behind it: The same natural endeavor of self-preservation that Spinoza articulates so transparently as an essence of our beings, the *conatus*, is called into action when we are confronted with the reality of suffering and especially the reality of death, actual or anticipated, our own or that of those we love. The very prospect of suffering and death breaks down the homeostatic process of the beholder. The natural endeavor for self-preservation and well-being responds to the breakdown with a struggle to prevent the inevitable and redress the balance. The struggle provokes us to find compensatory strategies for the homeodynamics now gone awry; and the awareness of the entire predicament is the cause of profound sorrow.

Once again, not every human being will react this way, for one reason or another, at one time or another. But for the many who do react in the manner I described, regardless of how effectively they manage to resolve the impasse and pull out of their darkness, there is a tragic dimension to this situation and it is exclusively human. How did this situation come to pass?

As far as I can fathom, the situation was a result, first, of having *feelings*—not just emotions but feelings—in particular the feelings of empathy with which we take full cognizance of our natural, emotive sympathy with the other; in the right circumstances empathy opens the door to sorrow. Second, the situation was a result of having two biological gifts, *consciousness* and *memory*, which we share with other species but that attain their greatest magnitude and degree of sophistication in humans. In the strict meaning of the word, consciousness signifies the presence of a mind with a self, but in practical human terms the word actually signifies more. With the help of autobiographical memory, consciousness provides us with a self enriched by the records of our own individual experience. When we face each new moment of life as conscious beings, we bring to bear on that moment the circumstances surrounding our past joys and sorrows, along with the imaginary circumstances of our anticipated future, those circumstances that are presumed to bring on more joys or more sorrows.

Were it not for this high level of human consciousness there would be no remarkable anguish to speak of, now or at the dawn of humanity. What we do not know cannot hurt us. If we had the gift of consciousness but were largely deprived of memory, there would be no remarkable anguish either. What we do know, in the present, but are unable to place in the context of our personal history, could only hurt us in the present. It is the two gifts combined, consciousness and memory, along with their abundance, that result in the human drama and confer upon that drama a tragic status, then and now. Fortunately, the same two gifts also are at the source of unbounded enjoyment, sheer human glory. Leading a life examined also brings a privilege and not just a curse. From this perspective, any project for human salvation—any project capable of turning a life examined into a life contented—must include ways to resist the anguish conjured up by suffering and death, cancel it, and substitute joy instead. The neurobiology of emotion and feeling tells us in suggestive terms that joy and its variants are preferable to sorrow and related affects, and more conducive to health and the creative flourishing of our beings. We should seek joy, by reasoned decree, regardless of how foolish and unrealistic the quest may look. If we do not exist under oppression or in famine and yet cannot convince ourselves how lucky we are to be alive, perhaps we are not trying hard enough.

Confronting death and suffering can forcefully disrupt the homeostatic state. Early humans may have first experienced this disruption once they acquired social emotions and feelings of empathy, emotions and feelings of joy and sorrow, extended consciousness with an autobiographical self, and the capacity to imagine entities and actions that could potentially alter the affective state and restore the homeostatic balance. (The first two conditions, emotions and feelings, social and not, already were budding in nonhuman species, as we have seen; the latter two, extended consciousness and imagination, were mostly new human gifts.) The yearning for homeostatic correctives would have begun as a response to anguish. Those individuals whose brains were capable of imagining such correctives and effectively restoring homeostatic balance would have been rewarded by longer life and larger progeny. Their genomic pattern would have had a better chance of dissemination and along with it the tendency for such responses would disseminate as well. The yearning and its beneficial consequences would have resurfaced again and again through generations. This is how a significant part of humanity might have incorporated both the conditions leading to personal sorrow and the seeking of compensatory comfort in its biological makeup.

Thus attempts at human salvation concern an accommodation to a death foretold or accommodations to physical pain and mental anguish. (For a time, of course, after the notion of immortality was invented, such attempts also were about preventing a life in hell.) There is a long history of such attempts. Intelligent individuals have been moved to create intriguing narratives that respond directly to the spectacle of tragedy, and aim at coping with the resulting distress by following religious percepts and practices. (This is not to suggest that the confrontation with death and suffering would have been the sole factor behind the development of religious narratives. The enforcement of ethical behaviors would have been another important factor and might have contributed just as much to the survival of individuals whose group succeeded in enforcing moral conventions.) Some of the well-known narratives promise postmortem rewards, and some promise comfort for the living, but the compensatory goal is the same. In a way, Spinoza is part of that historical response. Having been brought up in a religious community, and having rejected the solution the community proposed for human salvation, he was obliged to find another. Both the *Tractatus* and *The Ethics*, after their refined analyses of what is, are works about what ought to be and how to achieve it. To a considerable extent, however, Spinoza's solution also is a break with history.

Spinoza's Solution

Spinoza's system does have a God but not a provident God conceived in the image of humans. God is the origin of all there is before our senses, and it *is* all there is, an uncaused and eternal substance with infinite attributes. For practical purposes God is nature and is most clearly manifest in living creatures. This is captured in an often quoted Spinozism, the expression *Deus sive Natura*—God or Nature.[1] God has not revealed himself to humans in the ways portrayed in the Bible. You cannot pray to Spinoza's God.

You need not be in fear of this God because he will never punish you. Nor should you work hard in the hope of getting rewards from him because none will come. The only thing you may fear is your own behavior. When you fail to be less than kind to others, you punish yourself, there and then, and deny yourself the opportunity to achieve inner peace and happiness, there and then. When you are loving to others there is a good chance of achieving inner peace and happiness, there and then. Thus a person's actions should not be aimed at pleasing God, but rather at acting in conformity with the nature of God. When you do so, some kind of happiness results and some kind of salvation is achieved. *Now.* Spinoza's salvation—*salus*—is about repeated occasions of a kind of happiness that cumulatively make for a healthy mental condition.[2]

Spinoza rejected the notion that the prospect of after-death rewards or punishments was a proper incentive for ethical behavior. In a telling letter he lamented the man whose behavior is so guided: "He is one of those who would follow after his own lusts, if he were not restrained by the fear of hell. He abstains from evil actions and fulfills God's commands like a slave against his will, and for his bondage he expects to be rewarded by God with gifts far more to his taste than Divine love, and great in proportion to his original dislike of virtue."[3]

Spinoza makes room for two different roads to salvation: one accessible to all, the other more arduous and accessible only to those with disciplined and educated intellects. The accessible road requires a virtuous life in a virtuous *civitas,* obedient to the rules of a democratic state and mindful of God's nature, somewhat indirectly, with the help of some of the Bible's wisdom. The second road requires all that is needed by the first and, in addition, intuitive access to understanding that Spinoza prized above all other intellectual instruments, and which is itself based on abundant knowledge and sustained reflection. (Spinoza regards intuition as the most sophisticated means of achieving knowledge—intuition is Spinoza's knowledge of the third kind. But intuition occurs only after we accumulate knowledge and use reason to analyze it.) Predictably, Spinoza thought nothing of the effort required to achieve the desired results: "How could it be that if salvation was ready at hand and within reach without much effort, it would be neglected by almost all? All that is excellent is as difficult to obtain as it is rare." [*The Ethics,* Part V, notes to Proposition 42.]

For the first kind of salvation, Spinoza rejects biblical narratives as God's revelation, but endorses the wisdom embodied in the historical figures of Moses and Christ. Spinoza saw the Bible as a repository of valuable knowledge regarding human conduct and civil organization.[4]

The second road to salvation assumes that the requirements of the first are properly met—a virtuous life assisted by a sociopolitical system whose laws help the individual with the task of being fair and charitable to others—but then it goes further. Spinoza asks for an acceptance of natural events as necessary, in keeping with scientific understanding. For example, death and the ensuing loss cannot be prevented; we should acquiesce. The Spinoza solution also asks the individual to attempt a break between the emotionally competent stimuli that can trigger negative emotions—passions such as fear, anger, jealousy, sadness—and the very mechanisms that enact emotion. Instead, the individual should substitute emotionally competent stimuli capable of triggering positive, nourishing emotions. To facilitate this goal Spinoza recommends the mental rehearsing of negative emotional stimuli as a way to build a tolerance for negative emotions and gradually acquire a knack for generating positive ones. This is, in effect, Spinoza as mental immunologist developing a vaccine capable of creating antipassion antibodies. There is a Stoic color to the entire exercise, although it must be noted that Spinoza criticized the Stoics for assuming that the control of the emotions could ever be complete. (He criticized Descartes, too, for the same reason.) Spinoza was tough enough for my taste but not Stoic enough, it appears.

Spinoza's solution hinges on the mind's power over the emotional process, which in turn depends on a discovery of the causes of negative emotions, and on knowledge of the mechanics of emotion. The individual must be aware of the fundamental separation between emotionally competent stimuli and the trigger mechanism of emotion so that he can substitute *reasoned* emotionally competent stimuli capable of producing the most positive feeling states. (To some extent, Freud's psychoanalytical project shared these objectives.) Today, the new understanding of the machinery of emotion and feeling makes Spinoza's goal all the more achievable. Finally, Spinoza's solution asks the individual to reflect on life, guided by knowledge and reason, in the perspective of eternity—of God or Nature—rather than in the perspective of the individual's immortality.

The results of this effort are complicated and difficult to tease apart. Freedom is one of the results, not of the kind usually contemplated in discussions of free will, but something far more radical: a reduction of dependencies on the object-emotional needs that enslave us. Another result is that we intuit the essences of the human condition. That intuition is comingled with a serene feeling whose ingredients include pleasure, joy, delight, but for which the words "blessedness" and "beatitude" seem the most appropriate given the transparent texture of the feeling (*The Ethics,* Part V, Propositions 32 and 36, and their notes). This "intellectual" feeling is synonymous with an intellectual form of love for God—*amor intellectualis Dei.*[5]

Goethe noted that this process offers love without asking for love back, and wondered about what could be more generous and less disinterested than this attitude. But Goethe was not quite accurate. The individual does get something back in the form of the most desirable kind of human freedom—Spinoza believed that an entity is free only when it exists solely by the lights of its nature and when it acts solely by its own determination. The individual also achieves the most desirable kind of joy in Spinoza's canon, a joy that is perhaps best conceived as pure feeling almost liberated for once, from its obligate body twin.

Not everyone has been as kind as Goethe in their assessment of Spinoza's solution, and some consider it a hopeless muddle.[6] But neither the sincerity of the effort nor the pains and struggles that provided the incentive for it are in question. The Malamud character I invoked in chapter 1 captured the very least that can be said about these passages of the *The Ethics:* "...he was out to make a free man of himself." Nor is it in doubt that Spinoza managed to bring together reason and affect in a modern way. Spinoza's strategy to arrive at the intuited freedom and beatitude requires factual knowledge and reason. It also is curious that

someone who thought demonstrations were the eyes of the mind
spent a good part of his life creating the best possible lenses, instruments that helped the mind see so many new facts. Spinoza
embraced discovery of nature and knowledge as part of the diet of
a thinking person. How intriguing to think that the lenses he so
skillfully polished and the microscopes into which they went were
means of seeing clearly and were thus, in a way, instruments of
salvation. And how fitting for the time: Spinoza's was the age in
which numerous optical and mechanical devices were developed
both to permit scientific discovery and to make the discovery
process a source of pleasure.[7]

The Effectiveness of a Solution

How true does Spinoza's solution ring today and how effective
does it appear to be? The verdict, now and in Spinoza's time,
seems mixed.

For some, Spinoza's solution is a superior means to render
life meaningful and to make human society tolerable. The aim of
Spinoza's solution is to return us to the relative independence
humans lost after we gained extended consciousness and autobiographical memory. His route is through the use of reason and
feeling. Reason lets us see the way, while feeling is the enforcer
of our determination to see. What I find attractive in Spinoza's
solution is the recognition of the advantages of joy and the rejection of sorrow and fear, and the determination to seek the former
and obliterate the latter. Spinoza affirms life and turns emotion
and feeling into the means for its nourishing, a nice mixture of
wisdom and scientific foresight. On the way to the horizon of life
it is up to the individual to live in such a manner that the perfection of joy can be achieved frequently and thus render life worth
living. And because the process is grounded in nature Spinoza's
solution is immediately compatible with the view of the universe

that science has been constructing for the past four hundred years.

In other ways, Spinoza's solution is problematic. I am uneasy with the implication that Spinoza's solution works best in isolated self-centeredness, away from human intimacy. I find his asceticism rather impractical today. Spinoza does not go as far as the Greek and Roman Stoics in his divestiture of life's trappings, but he gets awfully close. We have been too corrupted not only by biting the apple of knowledge but also by swallowing it whole, and it seems unrealistic to divest ourselves of the baggage of things, facts, and habits that pervade our western high-tech life. Besides why should we? Why should Aristotle's wisdom not prevail here? Aristotle insisted that the contented life is a virtuous and happy life, but that health, wealth, love, and friendship are part of the contentment. I also am less than enthusiastic about the outward passivity of Spinoza's solution—never mind how inwardly active his beatitude may be. Others worry that arriving at the horizon of life Spinoza's solution merely offers death. There is no release from all the suffering and inequity that biology and society regularly visit upon humans, let alone compensation for the losses incurred along the way. Spinoza's God is an idea, rather than the flesh and blood that, for example, the Christian narrative has created. Spinoza might have been drunk with God, as Novalis said of him, but his God is rather dry.

For all the courage, perseverance, sacrifices, and discipline necessary to achieve that perfect joy, all one gets are moments of perfection. These are furtive glimpses of what? The divine? The balm is brief and one is left waiting for the next such moment, the next such glimpse. Depending on who you are, this is either bountiful or not nearly enough. But the fact that it can be seen as neither satisfying nor comfortable, let alone convenient, does not make it any less realistic.

If you ask of Spinoza's perspective, Hamlet's disquieting, inaugural question, "Who's there?"—meaning who is out there to let us persist as our endeavor of self-preservation mandates—the answer is unequivocal. No one. Aloneness is the stark reality, of Christ on the cross and Spinoza in the crushed pillows of his deathbed. And yet, Spinoza conjures up a means to elude that reality, a noble illusion meant to let us face the music and dance.

At the beginning of this book, I described Spinoza as both brilliant and exasperating. The reasons why I regard him as brilliant are obvious. But one reason why I find him exasperating is the tranquil certainty with which he faces a conflict that most of humanity has not yet resolved: the conflict between the view that suffering and death are natural biological phenomena that we should accept with equanimity—few educated human beings can fail to see the wisdom of doing so—and the no less natural inclination of the human mind to clash with that wisdom and feel dissatisfied by it. A wound remains and I wish it were not so. You see, I do prefer happy endings.

Spinozism

Intolerable as it was in his own time, Spinoza's brand of secular religiosity has been rediscovered or reinvented in the twentieth century. Einstein, for example, thought about God and religion in similar ways. He described the God of "the naive man" as "a being from whose care one hopes to benefit and whose punishment one fears; a sublimation of a feeling similar to that of a child for its father, a being to whom one stands to some extent in a personal relation, however deeply it may be tinged with awe."[8]

In describing his own religious feeling—the religious feeling of the "profounder sort of scientific minds"—Einstein wrote that such a feeling "...takes the form of a rapturous amazement at the harmony of natural law, which reveals an intelligence of

such superiority that, compared with it, all the systematic thinking and acting of human beings is an utterly insignificant reflection."9 In words of great beauty, Einstein described this feeling as
"...a sort of intoxicated joy and amazement at the beauty and
grandeur of this world, of which man can form just a faint notion.
This joy is the feeling from which true scientific research draws
its spiritual sustenance, but which also seems to find expression
in the song of birds." I believe this feeling, which Einstein called
cosmic, is a relative of Spinoza's *amor intellectualis Dei*, although
the two can be distinguished. Einstein's cosmic feeling is exuber-
ant, a mixture of heart-stopping awe and heart-beating-fast preparation for bodily communion with the world. Spinoza's *amor* is
more restrained. The communion is interior. Einstein seemed to
blend the two. He believed that the cosmic feeling is a hallmark of
religious geniuses of all ages, but that it never formed the basis
for any church. "Hence it is precisely among the heretics of every
age that we find men who were filled with the highest kind of religious feeling and were in many cases regarded by their contemporaries as atheists, sometimes also as saints. Looked at in this
light, men like Democritus, Francis of Assisi, and Spinoza are
closely akin to one another."10

William James's thinking on these matters also revealed a kinship with Spinoza's. This may be surprising given the near abyss
of time, place, and historical context that separated the two men.
Predictably, James's relation to Spinoza was not one of full acceptance. We learn from R. W. B. Lewis's biography that James first
read Spinoza in 1888 to teach a new course on the philosophy of
religion at Harvard University. The course eventually formed the
basis for James's *The Varieties of Religious Experience*.11 James resisted Spinoza on several issues. He did not endorse Spinoza's
provocative claim that "I will analyze the actions and appetites of
men as if it were a question of lines, of planes, and of solids."

Such "cold-blooded assimilations" are not to the liking of the adorable genius from Cambridge.[12] He also resisted what he diagnosed as Spinoza's sunny enthusiasm for life, his "healthy-mindedness."[13] His reason is fascinating. James divided human beings into two kinds: those with cheerful souls, and those with sick souls. The cheerful have a natural way of not seeing the tragedy of death, the horror of nature at its predatory worst, or the darkness in the recesses of the human mind. Irritatingly, for James, Spinoza appeared to be a cheerful soul, one of those born with "a constitutional incapacity for prolonged suffering," and with "a tendency to see things optimistically." For the Spinozas of this world, James said, "Evil is a disease; and worry over disease is itself an additional form of disease, which only adds to the original complaint."[14] Their optimism is natural.

James, on the other hand, was a "sick soul." Sick souls cannot contemplate nature and enjoy the spectacle, at least not all of the time, because the spectacle often is confirmedly horrible and unjust. One does not need to be a depressive to look at the world as a sick soul, although James did have a mood disorder—the magnificent development of the *Varieties* occurred on the rebound from a severe bout of depression. Curiously, however, James considers the sickness "good." Although the sickness is to be avoided in its major, pathological form, it should be present, to some degree, to force humans to confront reality without the misleading screen the sunny souls systematically interpose. Some dose of pessimism is good.

The casting of the problem of human salvation in cognitive and affective terms showed James at his most intellectually penetrating. It should be said, however, that he greatly exaggerated Spinoza's bubbliness. I do not believe Spinoza had any difficulty in seeing the darkness in nature, having experienced its effects himself. Quite the contrary. But he refused to *accept* darkness and to let it dominate the individual as a bad passion. He saw darkness as a part of existence and prescribed ways with which it can be minimized. Spinoza was resilient and courageous rather than naturally cheerful. He *strived* to be cheerful. He worked hard at canceling the feelings of fear and sorrow that nature inspires, with feelings of joy based on the discovery of nature. That discovery, almost perversely, included nature's cruelty and indifference.

Once James's resistances are overcome, however, there is much on his path to salvation that resembles Spinoza's. In both cases their experience of God was private. Both rejected the need for public rituals and congregations to have the experience of the divine. In fact, the arguments in James's sweeping dismissal of organized religion are quite Spinozian. Both James and Spinoza described the experience of the divine as pure feeling, a pleasurable feeling that is a source of completion, meaning, and enthusiasm for life. In the end, the important difference between the two is the baseline from which the healthy, salvatory feelings depart and can be measured. In Spinoza, the feeling of the divine rode on top of a baseline of reasoned equanimity regarding the world; in James the feeling of the divine started from a depressed baseline and it often pulled him up from the doldrums of his negative appraisal of nature. Otherwise, both James and Spinoza found God *within*, and James, using the budding knowledge of late nineteenth-century psychology which he himself helped construct, located the source of the divine not just within us but in the unconscious within us. He spoke of the religious experience as something "more," but told us that the "more" with which we can project ourselves "further" is in fact "hither."

Spinoza and James are pointing us toward a fruitful adaptation in the form of a natural life of the spirit. Their God is therapeutic in the sense that it restores the homeodynamic balance lost as a result of anguish. But neither man expected his God to listen.

Both believed that the restoration of balance is an individual and internal task, something to be achieved when sophisticated thinking and reasoning provoke the appropriate emotion and feeling. Both rationalized the process by acknowledging that human beings are mere occasions of subjective individuality in a largely mysterious universe. Neither could decipher the deepest rhymes and reasons of that universe.

Happy Endings?

How can we make our way toward a happy ending in a universe in which even the cheerful, sunny souls can so easily see human suffering in all its varieties from the unavoidable to the preventable? There are many that already have the answer, in the form of either a deeply felt religious faith or a protective insulation against sorrow of any kind. But what about the others, those who have neither resource? The honest answer is, of course, that I do not know, and that it would be presumptuous to offer a prescription for the happy endings of anyone else's life. But I can offer a word about my own view.

One path toward the happy ending I wish for comes from combining some features of Spinoza's contemplation with a more active stance aimed at the world around us. This path includes a life of the spirit that seeks understanding with enthusiasm and some sort of discipline as a source of joy—where understanding is derived from scientific knowledge, aesthetic experience or both. The practice of this life also assumes a combative attitude based on the belief that part of humanity's tragic condition can be alleviated, and that doing something about the human predicament is our responsibility. One benefit of scientific progress is the means to plan intelligent actions that can assuage suffering. Science can be combined with the best of a humanist tradition to permit a new approach to human affairs and lead to human flourishing.

To clarify this view let me begin by explaining what I mean by a life of the spirit. A friend of mine who follows the developments of biology with keen interest and is an equally avid seeker of the spiritual in life often asks me if the spirit can be defined and located in neurobiological terms. "What is the spirit?" "Where is it?" How can I answer? I must confess I do not favor the attempt to neurologize religious experiences, especially when the attempts take the form of identifying a brain center for God or justifying God and religion by finding their correlates in brain scans.[15] Yet, spiritual experiences, religious or otherwise, are mental processes. They are biological processes of the highest level of complexity. They occur in the brain of a given organism in certain circumstances and there is no reason why we should shy away from describing those processes in neurobiological terms provided we are aware of the limitations of the exercise. So, here are the answers to my friend's questions.

First, I assimilate the notion of spiritual to an intense experience of harmony, to the sense that the organism is functioning with the greatest possible perfection. The experience unfolds in association with the desire to act toward others with kindness and generosity. Thus to have a spiritual experience is to hold sustained feelings of a particular kind dominated by some variant of joy, however serene. The center of mass of the feelings I call spiritual is located at an intersection of experiences: Sheer beauty is one. The other is anticipation of actions conducted in "a temper of peace" and with "a preponderance of loving affections" (the quotations are James's but the concepts are Spinozian). These experiences can reverberate and become self-sustaining for brief periods of time. Conceived in this manner, the spiritual is an index of the organizing scheme behind a life that is well-balanced, well-tempered, and well-intended. One might venture that perhaps the spiritual is a partial revelation of the ongoing impulse behind life in some state of perfection. If feelings, as I suggested earlier in the book, testify to the state of the life process, spiritual feelings dig beneath that testimony, deeper into the substance of living. They form the basis for an intuition of the life process.[16]

Second, spiritual experiences are humanly nourishing. I believe that Spinoza was entirely on the mark in his view that joy and its variants lead to greater functional perfection. The current scientific knowledge regarding joy supports the notion that it should be actively sought because it does contribute to flourishing; likewise, that sorrow and related affects should be avoided because they are unhealthy. This entails the observance of a certain range of social norms—the recent evidence, presented in Chapter Four, that cooperative human behavior engages pleasure/reward systems in the brain supports this wisdom. Violation of social norms causes guilt or shame or grief, all of which are variants of unhealthy sorrow.

Third, we have the power to evoke spiritual experiences. Prayer and rituals, in the context of a religious narrative, are meant to produce spiritual experiences but there are other sources. It is often said that the secularity and crass commercialism of our age have made the spiritual all the more difficult to attain, as if the means to induce the spiritual were missing or becoming scarce. I believe this is not entirely true. We live surrounded by stimuli capable of evoking spirituality, although their saliency and effectiveness are diminished by the clutter of our environments and a lack of systematic frameworks within which their action can be effective. The contemplation of nature, the reflection on scientific discovery, and the experience of great art can be, in the appropriate context, effective emotionally competent stimuli behind the spiritual. Think of how listening to Bach, Mozart, Schubert, or Mahler can take us there, almost easily. This is an opportunity to generate positive emotions where negative emotions would otherwise arise, in the manner Spinoza recommended. It is clear, however, that the sort of spiritual experiences to which I am alluding are not equivalent to a religion. They lack the framework, as a result of which they also lack the sweep and the grandeur that attracts so many human beings to organized religion. Ceremonial rites and shared assembly do create ranges of spiritual experience different from those of the private variety.

Let us now turn to the delicate issue of "locating" the spiritual in the human organism. I do not believe that there is a brain center for spirituality in the good old phrenological tradition. But we can provide an account of how the process of arriving at a spiritual state may be carried out neurobiologically. Since the spiritual is a particular kind of feeling state, I see it as depending, neurally speaking, on the structures and operations outlined in Chapter Three, and especially on the network of somatosensing brain regions. The spiritual is a particular state of the *organism,* a delicate combination of certain body configurations and certain mental configurations. Sustaining such states depends on a wealth of thoughts about the condition of the self and the condition of other selves, about past and future, about both concrete and abstract conceptions of our nature.

By connecting spiritual experiences to the neurobiology of feelings, my purpose is not to reduce the sublime to the mechanic and by so doing reduce its dignity. The purpose is to suggest that the sublimity of the spiritual is embodied in the sublimity of biology and that we can begin to understand the process in biological terms. As for the results of the process, there is no need and no value to explaining them: The experience of the spiritual amply suffices.

Accounting for the physiological process behind the spiritual does not explain the mystery of the life process to which that particular feeling is connected. It reveals the connection to the mystery but not the mystery itself. Spinoza and those thinkers whose ideas have Spinozian elements make feelings come full circle, from life in progress, which is where they originate, to the sources of life, toward which they point.

I said the life of the spirit needs the complement of a combative stance. What does this mean? Seen in objective terms nature is neither cruel nor benign, but our practical view can be justifiably subjective and personal. On that view modern biology is now revealing that nature is even more cruel and indifferent than we previously thought. While humans are equal opportunity victims of nature's casual, unpremeditated evil, we are not obliged to accept it without response. We can try to find means to counteract the seeming cruelty and indifference. Nature lacks a plan for human flourishing, but nature's humans are allowed to devise such a plan. A combative stance, more so perhaps than the noble illusion of Spinoza's blessedness, seems to hold the promise that we shall never feel alone as long as our concern is the well-being of others.

And this is where I can answer the question posed at the beginning of the chapter: Knowing about emotion, feeling, and their workings does matter to how we live. At the personal level, this is quite certain. Within the next two decades, perhaps sooner, the neurobiology of emotion and feelings will allow biomedical science to develop effective treatments for pain and depression grounded on a sweeping understanding of how genes are expressed in particular brain regions and how these regions cooperate to make us emote and feel. The new treatments will aim at correcting specific impairments of a normal process rather than merely attacking symptoms in a general way. Combined with psychological interventions, the novel therapies will revolutionize mental health. The treatments available today will appear by then as gross and archaic as surgery without anesthesia appears to us now.

At the level of society the new knowledge is relevant as well. The relation between homeostasis and the governance of social and personal life discussed earlier should be helpful here. Some of the regulatory devices available to humans have been perfected through millions of years of biological evolution, as is the case of the appetites and emotions. Others have existed for just a few thousand years, as with the codified systems of justice and socio-political organization. Some are as good as they will ever get, set in genomic stone, not immutable to be sure, but as firm as biology can be. Some are a work in progress, a cauldron of tentative procedures aimed at the betterment of human affairs, but nowhere near the stability necessary to a harmonious life balance for all. And therein lies our opportunity to intervene and improve the human lot.

I am not suggesting that we attempt to manage social affairs with the same efficiency with which our brain maintains the basics of life. It probably cannot be done. Our goals should be more realistic. Besides, the repeated failures of such past and present attempts make us justifiably prone to cynicism. In fact, the temptation to recoil from any concerted effort to manage human affairs and to announce the end of the future is a comprehensible attitude. But nothing can guarantee defeat more certainly than retreating into isolated self-preservation. Much as it may sound naïve and utopian, especially after reading the morning paper or watching the evening news, there simply is no alternative to believing we can make a difference. There are some grounds for holding that belief. For example, the managing of specific problems such as drug addiction and violence will have a greater chance of success if informed by a new scientific understanding

of the human mind, including the very knowledge of life regulation that emerges from the science of emotion and feeling. The
same is likely to apply to a broad range of social policies. No doubt
the failure of past social engineering experiments is due, in some
part, to the sheer folly of the plans or the corruption of their execution. But the failure also may have been due to the misconceptions of the human mind that informed the attempts. Among
other negative consequences, the misconceptions resulted in a
demand for human sacrifices that most humans find difficult or
impossible to achieve; in an ignorant disregard for the aspects of
biological regulation that are now becoming scientifically transparent and that Spinoza intuited in the *conatus;* and in a blindness
to the dark side of social emotions that finds expression in tribal-
ism, racism, tyranny, and religious fanaticism. But that is the
past. Now we are forewarned, and entitled to a new beginning.

I believe the new knowledge may change the human playing
field. And this is why, all things considered, in the middle of
much sorrow and some joy, we can have hope, an affect for which
Spinoza, in all his bravery, did not have as much regard as we
common mortals must. He defined it as follows: "Hope is nothing else but an inconstant joy, arising from the image of something future or past, whose outcome to some extent we doubt."[17]

Appendix I

Before, During, and After Spinoza's Time

1543   Death of Copernicus (born 1473), who proposed that the earth revolves around the sun and not the other way around.

1546   Death of Martin Luther (born 1483), who was excommunicated by the Catholic Church in 1521; founded the Lutheran Church.

1564   Birth of Galileo Galilee, William Shakespeare, and Christopher Marlowe.

Death of Jean Calvin, who founded Calvinism (the Presbyterian Church today) in 1536.

1572   Luis de Camões publishes *The Lusiad*.

1588   Birth of Thomas Hobbes, the English philosopher who took a clearly materialistic view of the mind. He had a significant influence on Spinoza.

1592   Death of Michel de Montaigne (born 1533), whose essays published in 1588 had a significant intellectual impact at the time.

1593   Christopher Marlowe dies in an accident.

1596 Birth of René Descartes.

1600 Giordano Bruno burned at the stake for siding with Copernicus and holding pantheistic beliefs.

1601 William Shakespeare's mature *Hamlet* is performed. The age of questioning begins.

1604 Shakespeare's *King Lear* performed.

Francis Bacon's *Advancement of Learning*.

Miguel de Cervantes's *Don Quixote* published.

1606 Birth of Rembrandt van Rijn.

1610 Galileo builds a telescope. His study of the stars leads him to adopt Copernicus's views on movements of the sun and earth.

1616 Shakespeare dies at fifty-two, still revising *Hamlet*.

Cervantes, sixty-nine, dies on the same day.

1629 Birth of Christiaan Huygens (d. 1695), astronomer and physicist. Intellectual peer, correspondent, sometime neighbor, and lens customer of Spinoza.

1632 Birth of John Locke.

Birth of Spinoza.

Rembrandt paints *The Anatomy Lesson of Dr. Tulp*.

1633 Galileo is convicted and placed under house arrest.

Descartes thinks twice about publishing views on human nature resulting from his research on human anatomy and physiology.

1633   William Harvey describes the circulation of blood.

1638   Birth of Louis XIV, who eventually reigns until 1715.

1640   Uriel da Costa, a Portuguese philosopher of Jewish origin,
       raised as a Catholic and later converted to Judaism, is first
       excommunicated and then reintegrated but physically pun-
       ished by the Portuguese Synagogue in Amsterdam. He
       commits suicide shortly thereafter but not before finishing
       his book, *Exemplar Vitae Humanae*.

1642   Death of Galileo.

       Birth of Isaac Newton (d. 1727).

1650   Death of Descartes.

1652   Death of Spinoza's father, Miguel de Espinoza.

1656   Spinoza is excommunicated by the Portuguese Synagogue
       and prevented from contact with any Jews, including fam-
       ily and friends. Thereafter he lives alone, in various Dutch
       cities, until 1670.

1670   Spinoza moves to The Hague.

       Anonymous publication of Spinoza's *Tractatus Politicus Re-
       ligiosus* in Latin.

1677   Death of Spinoza.

       Near anonymous publication of Spinoza's *Opera Posthuma*
       in Latin. Collection includes *The Ethics*.

1678   Publication of the body of Spinoza's work in Dutch and
       French. Secular and ecclesiastical authorities enforce pro-
       hibition of Spinoza's books throughout Europe. His work
       circulates illegally.

1684 John Locke's exile in Holland to 1689.

1687 Publication of Newton's treatise on gravitation.

1690 Locke publishes *Essays Concerning Human Understanding* and *Two Treatises on Government* at age sixty.

1704 Locke dies at age seventy-two.

1743 Birth of Thomas Jefferson.

1748 Montesquieu publishes *L'Esprit des Lois.*

1764 Voltaire's *Philosophical Dictionary* is published five years after his *Candide.*

1772 Conclusion of the publication of the *Encyclopédie,* the centerpiece work of the Enlightenment, under the direction of Denis Diderot and Jean-le-Rond d'Alembert.

1776 Jefferson writes the Declaration of Independence.

1789 The French Revolution.

1791 The First Amendment to the United States Constitution.

Appendix II

Brain Anatomy

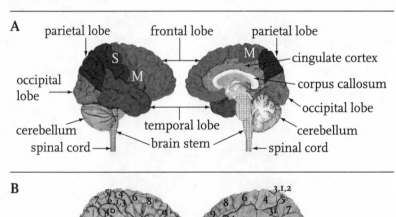

A

parietal lobe   frontal lobe   parietal lobe

occipital lobe

S

M

M

cingulate cortex

corpus callosum

occipital lobe

cerebellum   temporal lobe   cerebellum

spinal cord   brain stem   spinal cord

B

Brodmann's areas

Figure 1. The two top panels (A) depict the externally visible divisions of the central nervous system: the cerebrum, with its four lobes (occipital, parietal, temporal, frontal) and the cingulate cortex; the cerebellum; the brain stem; and the spinal cord. The left panel shows the lateral (external) view of the right cerebral hemisphere. The right panel shows the medial (internal) view of the same right cerebral hemisphere. S = sensory; M = motor.

The bottom panels (B) show the same lateral and medial views of the right hemisphere, but the cerebral cortex is now divided according to Brodmann's cytoarchitectural regions: each number corresponds to a part of the cerebral cortex recognizable by its distinctive cellular architecture. The distinctive architecture is due to the fact that the types of neurons and their layering differ from area to area, and that the neuron "projections" each area receives from other parts of the brain and sends to other parts of the brain also is different. The diverse architecture and the strikingly different inputs and outputs of each area explain why each area operates so differently and contributes so uniquely to the functions of the ensemble.

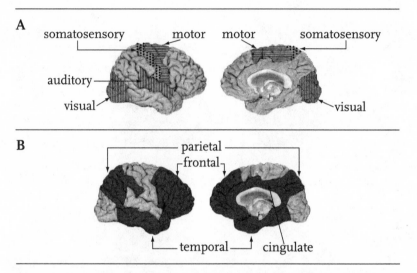

Figure 2. Two types of cerebral cortex. The top panels (A) depict the motor cortices and the primary (so-called "early") sensory cortices for vision, hearing and body sensations (somatosensory). The cortex of the insula, which is also related to body sensations, is not visible because it is hidden by the lateral parietal and frontal cortices (see Figure 3). The shadowed regions in B cover the association cortices of the several lobes and of the cingulate region. These cortices are also known as "higher-order" and "integrative."

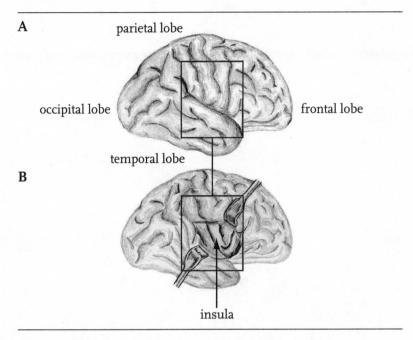

Figure 3. A depiction of the insula, a critical component of the somato-sensory cortex, which is visible only when the overlying cortices (seen in A) are retracted (as shown in B).

Notes

CHAPTER 1: Enter Feelings

1   The structure and the operation of the nervous system of a living being can be studied at different levels of organization, from the small and simple (the microscopic molecules that constitute an enzyme or a neurotransmitter) to the large and complex (the systems of macroscopic brain regions and their interconnections on the basis of whose operation we behave and think). Most of the work discussed in this book focuses on the latter level—the large-scale system level. The ultimate goal of our efforts is to connect the evidence from that level to evidence at levels below and above. The levels below include circuits and pathways; cells and chemical signal transmission. The levels above include mental and social phenomena.

In spite of the major importance of certain regions in the unfolding of this or that phenomenon, the processes of mind and behavior result from the concerted operation of the *many* regions that constitute brain systems, small and large. None of the grand functions of the human mind—perception, learning and memory, emotion and feeling, attention, reasoning, language, motion—arises in a *single* center of the brain. Phrenology, the idea that one brain center would produce one grand mental ability is a thing of the past. It is appropriate, however, to recognize that brain regions are highly specialized in terms of the contribution they can make to the overall function of a system. Their contribution is both special and flexible, subject to the vagaries of the occasion and to global influences, a bit like the string player in the symphony orchestra who will play well or not depending on his colleagues, the conductor, his mood, and so forth.

In addition to modern imaging scanners, which allow us to investigate brain anatomy and function, there are many other ways to probe the brain, ranging from the study of the electrical and magnetic phenomena produced by brain activity, to the study of gene expression in small brain regions.

2   Yakov tells the Magistrate what Spinoza means to him. Bernard Malamud, *The Fixer* (New York: Farrar, Straus and Giroux, 1966/ Viking Penguin, 1993).

3   Spinoza, *The Ethics,* Part III (New York: Dover Press, 1955). Other editions of *The Ethics* used in the text include Edwin Curley's in *The Collected Works of Spinoza* (Princeton University Press, 1985); and Joaquim de Carvalho's *Ética* (Relogio e Água, Lisbon, 1992).

4   Spinoza, *The Ethics,* Part IV, Proposition 7, ibid.

5   Spinoza, *The Ethics,* Part I, ibid.

6   Spinoza, *The Ethics,* Part II, ibid.

7   Jean Pierre Changeux is a notable exception. He closes his 1983 book *L'Homme Neuronal* with a Spinoza quote. Jean Pierre Changeux, *Neuronal Man: The Biology of Mind* (New York: Pantheon, 1985). He also discusses the relevance of Spinoza to neuroscience in *La Nature et La Règle* (Paris: Editions Odile Jacob, 1998) with Paul Ricoeur. Other thinkers who have noted a connection between Spinoza and modern psychology or biology are Stuart Hampshire, *Spinoza* (New York: Penguin Books, 1951); Errol Harris, *The Foundations of Metaphysical Science* (New York: Humanities Press, 1965); Edwin Curley, *Behind the Geometrical Method: A Reading of Spinoza's Ethics* (Princeton, N.J.: Princeton University Press, 1988).

8   Jonathan Israel makes a powerful argument for Spinoza's behind-the-scenes role in the Enlightenment in *Radical Enlightenment: Philosophy and the Making of Modernity* (New York: Oxford University Press, 2001). Also see chapter 6 of this volume for comments on Spinoza's role in the Enlightenment.

9   Gilles Deleuze, *Spinoza: A Practical Philosophy* (San Francisco: City Lights Books, 1988); Michael Hardt, A. Negri, *Empire* (Cambridge, Mass.: Harvard University Press, 2000); Henri Atlan, *La Science est-elle inhumaine?* (Paris: Bayard, 2002).

10 Spinoza, *A theologico-political treatise and A political treatise* (R. H. M. Elwes, *Benedict de Spinoza: A theologico-political treatise and a Political Treatise,* New York: Dover Publications, 1951).

11 Simon Schama, *An Embarrassment of Riches* (New York: Random House, 1987).

12 Descartes apparently had used the quote in life. It comes from Ovid's *Tristra*: "Bene qui latuit, bene vixit."

CHAPTER 2: Of Appetites and Emotions

1 Shakespeare, *Richard II*. Act 4, Scene 1.

2 The use of "mind" and "body" is not an inadvertent slip into substance dualism of the Cartesian variety. As explained in Chapter Five, although I see the phenomena we usually call "mind" and "body" as arising from a single biological "substance," I treat mind and body as different research objects for the same reasons that bring me to distinguish emotion and feeling: a research strategy aimed at advancing the understanding of the integrated whole formed by mind-body or emotion-feeling.

3 In his writings on this subject Spinoza uses neither the word emotion nor the word feeling but rather affect—in Latin, *affectus*—a word that is appropriate for both concepts. He says that, "By *affectus* I mean the modifications of the body, whereby the active power of the said body is increased or diminished, aided or constrained, and also the ideas of such modifications" (Spinoza, *The Ethics*, Part III). When he wishes to clarify his precise meaning he qualifies affect and lets us know if he means the largely external or the exclusively internal aspect of the phenomenon, the emotion or the feeling. I suspect he would welcome the distinction I am proposing because the distinction is founded on the identification of different events in the process of "being affected," precisely as Spinoza's parallel terms *appetite* and *desire* are.

Of related interest: one of the most widely used English translations of Spinoza's works—by R. H. M. Elwes, published in England in 1883—renders the Latin *affectus* as emotion and helps perpetuate the incorrect usage of these terms. Edwin Curley's modern American translation appropriately renders *affectus* as affects. To help complicate matters Elwes renders Spinoza's *laetitia* and *tristitia* as pleasure and pain, where a more acceptable translation is happiness/joy and sadness/sorrow.

4   Hines to Buck Mulligan speaking of Stephen Dedalus. Part II.
     James Joyce, *Ulysses* (New York: Random House, 1986).

5   The word homeodynamics is even more appropriate than ho-
     meostasis because it suggests the process of seeking an adjustment
     rather than a fixed point of balance. Steven Rose introduced the
     term homeodynamics for these same reasons (Steven Rose, *Life-
     lines: Biology Beyond Determinism*, New York: Oxford University
     Press, 1998).

6   Ross Buck, "Prime theory: An integrated view of motivation and
     emotion," *Psychological Review* 92 (1985): 389–413; Ross Buck, "The
     biological affects: A typology," *Psychological Review*.

7   See Paul Griffiths, *What Emotions Really Are* (Chicago: University of
     Chicago Press, 1997), for a discussion on the problems of classify-
     ing emotions. The distinction between emotions-proper and the
     other bioregulatory reactions is not sharp. In general, emotions-
     proper are triggered by many objects and events with certain shared
     characteristics, rather than by one specific object or event, and the
     triggering process tends to be more complex. Also, the triggering
     stimulus is virtually always external in the case of emotions-proper
     and internal in the case of other reactions.

8   Monica S. Moore, Jim DeZazzo, Alvin Y. Luk, Tim Tully, Carol M.
     Singh, Ulrike Heberlein, "Ethanol intoxication in Drosophila: Ge-
     netic and pharmacological evidence for regulation by the cAMP sig-
     naling pathway," *Cell* 93 (1998): 997–1007.

9   Ralph J. Greenspan, Giulio Tononi, Chiara Cirelli, Paul J. Shaw,
     "Sleep and the fruit fly," *Trends in Neurosciences* 24 (2001): 142–45.

10   Irving Kupfermann, Vincent Castellucci, Harold Pinsker, Eric Kan-
     del, "Neuronal correlates of habituation and dishabituation of the
     gill-withdrawal reflex in Aplysia," *Science* 167 (1970): 1743–45.

11   Antonio Damasio, *Descartes' Error: Emotion, Reason, and the Human
     Brain* (New York: Grosset/Putnam, 1994; HarperCollins, 1995). To
     some extent Daniel Stern's notion of vitality affects is coextensive
     with the concept of background emotions. Daniel N. Stern, *The In-
     terpersonal World of the Infant* (New York: Basic Books, 1985).

12   Paul Ekman, "An argument for basic emotions," *Cognition and
     Emotion* 6 (1992): 169–200. Charles Darwin, *The Expression of the
     Emotions in Man and Animals* (New York: New York Philosophical
     Library, 1872).

13 Jaak Panksepp, *Affective Neuroscience: The Foundations of Human and Emotions* (New York: Oxford University Press, 1998); Richard Davidson, "Prolegomenon to emotion: Gleanings from neuropsychology," *Cognition and Emotion* 6 (1992): 245–68; Richard Davidson, William Irwin, "The functional neuroanatomy of emotion and affective style," *Trends in Cognitive Sciences* 3 (1999): 211–21; Raymond Dolan, Paul Fletcher, J. Morris, N. Kapur, J. F. Deakin, Christopher D. Frith, "Neural activation during covert processing of positive emotional facial expressions," *NeuroImage* 4 (1996): 194–200; Joseph LeDoux, *The Emotional Brain: The Mysterious Underpinnings of Emotional Life* (New York: Simon and Schuster, 1996); Michael Davis and Y. Lee, "Fear and anxiety: possible roles of the amygdala and bed nucleus of the stria terminalis," *Cognition and Emotion* 12 (1998): 277–305; Edmund Rolls, *The Brain and Emotion* (New York: Oxford University Press, 1999); Ralph Adolphs, Daniel Tranel, Antonio Damasio, "Impaired recognition of emotion in facial expressions following bilateral damage to the human amygdala," *Nature* 372 (1994): 669–72; Ralph Adolphs, Daniel Tranel, Antonio R. Damasio, "The human amygdala in social judgment," *Nature* 393 (1998): 470–74; Ralph Adolphs, "Social cognition and the human brain," *Trends in Cognitive Sciences* 3 (1999): 469–79; Ralph Adolphs, Hanna Damasio, Daniel Tranel, Gregory Cooper, Antonio Damasio, "A role for somatosensory cortices in the visual recognition of emotion as revealed by 3-D lesion mapping," *The Journal of Neuroscience* 20 (2000): 2683–90; Ralph Adolphs, "Neural mechanisms for recognizing emotion," *Current Opinion in Neurobiology* 12 (2002): 169–78; Jean-Didier Vincent, *Biologie des Passions* (Paris: Editions Odile Jacob, 1986); Nico Frijda, *The Emotions* (Cambridge, U.K., New York: Cambridge University Press, 1986); Karl Pribram, *Languages of the Brain: Experimental Paradoxes and Principles in Neuropsychology* (Englewood Cliffs, N.J.: Prentice-Hall, 1971). Stephen W. Porges, "Emotion: An evolutionary byproduct of the neural regulation of the autonomic nervous system," *Annals of the New York Academy of Sciences* 807 (1997): 62–77.

14 Paul Rozin, L. Lower, R. Ebert, "Varieties of disgust faces and the structure of disgust," *Journal of Personality & Social Psychology* 66 (1994): 870–81.

15 Richard Davidson and W. Irwin (cited earlier); Raymond Dolan, et al. (cited earlier); Helen Mayberg, Mario Liotti, Steven K. Brannan,

Scott McGinnis, Roderick K. Mahurin, Paul A. Jerabek, J. Arturo Silva, Janet L. Tekell, C. C. Martin, Jack L. Lancaster, Peter T. Fox, "Reciprocal limbic-cortical function and negative mood: Converging PET findings in depression and normal sadness," *American Journal of Psychiatry* 156 (1999): 675–82; Richard Lane, Eric M. Reiman, Geoffry L. Ahern, Gary E. Schwartz, Richard J. Davidson, "Neuroanatomical correlates of happiness, sadness, and disgust," *American Journal of Psychiatry* 154 (1997): 926–33; Wayne Drevets, Joseph L. Price, Joseph R. Simpson Jr., Richard D. Todd, Theodore Reich, Michael Vannier, Marcus E. Raichle, "Subgenual prefrontal cortex abnormalities in mood disorders," *Nature* 386 (1997): 824–27.

16   Frans de Waal, *Good Natured* (Cambridge: Harvard University Press, 1997); Hans Kummer, *The Quest of the Sacred Baboon* (Princeton, N.J.: Princeton University Press, 1995); Berud Heinrich, *The Mind of the Raven* (New York: HarperCollins, 1999); Marc D. Hauser, *Wild Minds* (New York: Henry Holt, 2000).

17   Robert Hinde, "Relations between levels of complexity in the behavioral sciences," *Journal of Nervous & Mental Disease* 177 (1989): 655–67.

18   Cornelia Bargmann, "From the nose to the brain," *Nature* 384 (1996): 512–13.

19   For a modern discussion of possible interactions between the world of affect and that of evolution, see Jaak Panksepp, *Affective Neuroscience: The Foundations of Human and Emotions* (cited earlier); and Mark Solms, *The Brain and the Inner World: An Introduction to the Neuroscience of Subjective Experience* (New York: Other Press, 2001).

20   Ross Buck (cited earlier).

21   Antonio Damasio, "Fundamental feelings," *Nature* 413 (2001): 781. The goal of this provisional definition is to be as specific and inclusive as possible while respecting the research-oriented separation between emotion and feeling I proposed to follow earlier. There are mental elements in this definition (the appraisal/presentation of an emotionally competent stimulus); neural and bodily physiological elements; an evolutionary perspective; and a statement of functional purpose. The definition avoids a restrictive point of view, e.g., defining emotions "as states elicited by rewards and punishments" in a framework in which "a reward is anything for which an animal will

work," and "a punishment is anything that an animal will work to avoid or escape," as proposed by E. T. Rolls in *Behavioral and Brain Sciences* 23 (2000): 177–234.

22 The weight of my discussion falls on the processes that unfold beyond the appraisal phase for the good reason that this is the least understood phase of the emotional reaction and one that promises to reveal neurobiological underpinnings for the feeling part of the cycle. Fortunately for us, the appraisal process is partly accessible to introspection and has been investigated in great detail, based on a wealth of human experience recorded not just on pages of philosophy and science, but also literature, as Martha Nussbaum has shown. As noted at the outset, the focus of my inquiry is the more proximate neurobiological mechanism for producing an emotion. (Martha Nussbaum, *Upheavals of Thought*, New York: Cambridge University Press, 2001.)

23 Studies focused on the amygdala reveal that a variety of glutamate receptor known as the NMDA receptor is a key in these processes, especially its NR2B subunit. For example, disruption of this subunit prevents fear conditioning; on the other hand, the same subunit can be manipulated genetically to produce enhancements of emotional learning. The NMDA receptor is also involved in the activation of an enzyme, cAMP dependent protein kinase, on which protein synthesis and new learning depend. See Eric Kandel, James Schwartz, Thomas Jessell, *Principles of Neural Science*, chapters on Learning and Memory, McGraw-Hill, 4th Edition (2002); J. LeDoux, *The Synaptic Self*, Simon and Schuster (2002).

24 Joseph LeDoux (cited earlier); Ralph Adolphs (cited earlier); Raymond Dolan (cited earlier); David Amaral, "The primate amygdala and the neurobiology of social behavior: implications for understanding social anxiety," *Biological Psychiatry* 51 (2002): 11–17; Lawrence Weiskrantz, "Behavioral changes associated with ablations of the amygdaloid complex in monkeys," *Journal of Comparative and Physiological Psychology* 49 (1956): 381–91.

25 Hiroyuki Oya, Hiroto Kawasaki, Matthew Howard, Ralph Adolphs, "Electrophysiological responses recorded in the human amygdala discriminate emotion categories of visual stimuli," *Journal of Neuroscience* (in press).

26 Paul J. Whalen, Scott L. Rauch, Nancy L. Etcoff, Sean C. McInerney, Michael B. Lee, Michael A. Jenike, "Masked presentations of emotional facial expressions modulate amygdala activity without explicit knowledge," *Journal of Neuroscience* 18 (1998): 411–18.

27 Arnie Ohman, Joaquim J. Soares, "Emotional conditioning to masked stimuli: expectancies for aversive outcomes following non-recognized fear-relevant stimuli," *Journal of Experimental Psychology: General* 127 (1998): 69–82; J. S. Morris, Arnie Ohman. Raymond J. Dolan, "Conscious and unconscious emotional learning in the human amygdala," *Nature* 393 (1998): 467–70.

28 Patrik Vuilleumier, S. Schwartz, "Modulation of visual perception by eye gaze direction in patients with spatial neglect and extinction," *NeuroReport* 12 (2001): 2101–4; Patrik Vuilleumier, S. Schwartz, "Beware and be aware: capture of spatial attention by fear-related stimuli in neglect," *NeuroReport* 12 (2001): 1119–22; Patrik Vuilleumier, S. Schwartz, "Emotional facial expressions capture attention," *Neurology* 56 (2001): 153–58; Beatrice de Gelder, Jean Vroomen, G. Pourtois, Lawrence Weiskrantz, "Non-conscious recognition of affect in the absence of striate cortex," *NeuroReport* 10 (1999): 3759–63.

29 Antonio Damasio, Daniel Tranel, Hanna Damasio, "Somatic markers and the guidance of behavior: Theory and preliminary testing," in H. S. Levin, H. M. Eisenberg, and A. L. Benton, eds., *Frontal Lobe Function and Dysfunction* (New York: Oxford University Press, 1991), 217–29; Antonio Damasio, "The somatic marker hypothesis and the possible functions of the prefrontal cortex," *Transactions of the Royal Society* (London) 351 (1996): 1413–20; Antoine Bechara, Antonio Damasio, Hanna Damasio, Steven Anderson, "Insensitivity to future consequences following damage to human prefrontal cortex," *Cognition* 50 (1994): 7–15; Antoine Bechara, Daniel Tranel, Hanna Damasio, Antonio Damasio, "Failure to respond autonomically to anticipated future outcomes following damage to prefrontal cortex," *Cerebral Cortex* 6 (1996): 215–25; Antoine Bechara, Hanna Damasio, Daniel Tranel, Antonio Damasio, "Deciding advantageously before knowing the advantageous strategy," *Science* 275 (1997): 1293–94.

30 Hiroto Kawasaki, Ralph Adolphs, Olaf Kaufman, Hanna Damasio, Antonio Damasio, Mark Granner, Hans Bakken, Tomokatsu Hori,

Matthew A. Howard, "Single-unit responses to emotional visual stimuli recorded in human ventral prefrontal cortex," *Nature Neuroscience* 4 (2001): 15–16.

31 Jaak Panksepp, *Affective Neuroscience: The Foundations of Human and Emotions* (cited earlier).

32 Paul Ekman, "Facial expressions of emotion: New findings, new questions," *Psychological Science* 3 (1992): 34–38.

33 Boulos-Paul Bejjani, Philippe Damier, Isabella Arnulf, Lionel Thivard, Anne-Marie Bonnet, Didier Dormont, Philippe Cornu, Bernard Pidoux, Yves Samson, Yves Agid, "Transient acute depression induced by high-frequency deep-brain stimulation," *New England Journal of Medicine* 340 (1999): 1476–80.

34 Itzhak Fried, Charles L. Wilson, Katherine A. MacDonald, Eric J. Behnke, "Electric current stimulates laughter," *Nature* 391 (1998): 650.

35 Antonio Damasio, *Descartes' Error: Emotion, Reason, and the Human Brain* (cited earlier). The original observations of these phenomenon go back to my mentor Norman Geschwind.

36 Josef Parvizi, Steven Anderson, Coleman Martin, Hanna Damasio, Antonio R. Damasio, "Pathological laughter and crying: a link to the cerebellum." *Brain* 124 (2001): 1708–19.

37 The cerebellum possibly adjusts laughter and crying behaviors to specific contexts, e.g., social situations in which such behaviors should be inhibited. The cerebellum may set the threshold at which the induction-effector apparatus responds to a stimulus thus producing, or not, laughter or crying. These modulatory cerebellar actions would occur automatically as a result of learning (i.e., pairing certain social contexts to certain profiles and levels of emotional response). The cerebellum can perform these modulatory actions for two reasons. First, because it receives signals from telencephalic structures that convey the cognitive/social context of a stimulus, allowing the computations performed by the cerebellum to take into account such contexts. Second, because the cerebellar projections to the brain stem and telencephalic inductor and effector sites allow the cerebellum to coordinate the responses whose ensemble constitutes laughter or crying. Those responses involve the coordination of facial, laryngopharyngeal, and rhythmic diaphragmatic movements.

See Schmahmann for a discussion of cerebellar circuitry and function relevant to these issues. Jeremy D. Schmahmann, Deepak N. Pandya, "Anatomic organization of the basilar pontine projections from prefrontal cortices in rhesus monkey," *Journal of Neuroscience* 17 (1997a): 438–58; Jeremy D. Schmahmann, Deepak N. Pandya, "The cerebrocerebellar system," *International Review Neurobiology* 41 (1997b): 31–60.

CHAPTER 3: Feelings

1   Suzanne Langer contributed powerful analyses of the phenomena of feeling in her books, e.g., *Philosophy in a New Key* (Harvard University Press, 1942); *Philosophical Sketches* (Johns Hopkins Press, 1962). Suzanne Langer and her mentor, Alfred North Whitehead, are kindred spirits on this issue, as is the philosopher Errol Harris with whose work I became acquainted in the last stages of preparing this book, on the advice of Samuel Attard. Errol E. Harris, *The Foundations of Metaphysical Science* (New York: Humanities Press, 1965).

2   My colleague David Rudrauf believes that resistance to variance is a main cause of our experience of emotion, an idea that accords well with Francisco Varela's overall conception of the organism in biophysical terms. On this hypothesis, part of what we feel would correspond to a resistance to the upheaval caused by emotion, to the tendency to control the ongoing emotive perturbation.

3   This is relevant to the issue of *qualia*, for those who worry about that much-debated problem, but this is not the occasion to discuss the issue. Suffice it to say that when feelings are discussed in the broader frame presented here, and when one realizes that hardly any perception slips by without producing an "emotive" perturbation, the notion of qualia becomes more transparent.

4   Thomas Insel, "A neurobiological basis of social attachment," *American Journal of Psychiatry* 154 (1997): 726–36.

5   For a modern and science-inspired treatment of the distinctions among sex, attachment, and romantic love see Carol Gilligan, *The Birth of Pleasure* (Knopf, 2002); Jean-Didier Vincent, *Biologie des Passions* (Paris: Editions Odile Jacob, 1994); Alain Prochiantz, *La Biologie dans le Boudoir* (Paris: Editions Odile Jacob, 1995). For the

classical view on the same subject one must turn to Gustave Flaubert, Stendal, James Joyce, and Marcel Proust.

6 Antonio R. Damasio, Thomas J. Grabowski, Antoine Bechara, Hanna Damasio, Laura L. B. Ponto, Josef Parvizi, Richard D. Hichwa, "Subcortical and cortical brain activity during the feeling of self-generated emotions," *Nature Neuroscience* 3 (2000): 1049–56.

7 Hugo D. Critchley, Christopher J. Mathias, Raymond J. Dolan, "Neuroanatomical basis for first- and second-order representations of bodily states," *Nature Neuroscience* 4 (2001): 207–12. Note on other functional imaging studies of emotion/feeling: Helen S. Mayberg, Mario Liotti, Steven K. Brannan, Scott McGinnis, Roderick K. Mahurin, Paul A. Jerabek, Arturo Silva, Janet L. Tekell, Clifford C. Martin, Jack L. Lancaster, Peter T. Fox, "Reciprocal limbic-cortical function and negative mood: Converging PET findings in depression and normal sadness," 1999, cited earlier: 675–82; Richard Lane, et al., "Neuroanatomical correlates of happiness, sadness, and disgust" (cited earlier); Wayne Drevets, et al., "Subgenual prefrontal cortex abnormalities in mood disorders" (cited earlier); Hugo D. Critchley, Rebecca Elliot, Christopher J. Mathias, Raymond J. Dolan, "Neural activity relating to generation and representation of galvanic skin conductance responses: A functional magnetic resonance imaging study," *Journal of Neuroscience* 20 (2000): 3033–40.

8 Dana M. Small, Robert J. Zatorre, Alain Dagher, Alan C. Evans, Marilyn Jones-Gotman, "Changes in brain activity related to eating chocolate: from pleasure to aversion," *Brain* 124 (2001): 1720–33; A. Bartels, Semir Zeki, "The neural basis of romantic love," *NeuroReport* 11 (2000): 3829–34; Lisa M. Shin, Darin D. Dougherty, Scott P. Orr, Roger K. Pitman, Mark Lasko, Michael L. Macklin, Nathaniel M. Alpert, Alan J. Fischman, Scott L. Rauch, "Activation of anterior paralimbic structures during guilt-related script-driven imagery," *Society of Biological Psychiatry* 48 (2000): 43–50; Sherif Karama, André Roch Lecours, Jean-Maxime Leroux, Pierre Bourgouin, Gilles Beaudoin, Sven Joubert, Mario Beauregard, "Areas of brain activation in males and females during viewing of erotic film excerpts," *Human Brain Mapping* 16 (2002): 1–13.

9 Jaak Panksepp, "The emotional sources of chills induced by music," *Music Perception* 13 (1995): 171–207.

10 Anne J. Blood, Robert J. Zatorre, "Intensely pleasurable responses to

music correlate with activity in brain regions implicated in reward and emotion," *Proceedings of the National Academy of Sciences* 98 (2001): 11818–23.

11 Abraham Goldstein, "Thrills in response to music and other stimuli," *Physiological Psychology* 3 (1980): 126–169. We know that the administration of naloxone, a substance that blocks the action of opioids, also suspends the experience of chills making it likely that opioids are the mediator for these feelings.

12 Kenneth L. Casey, "Concepts of pain mechanisms: the contribution of functional imaging of the human brain," *Progress in Brain Research* 129 (2000): 277–87.

13 In a related experiment Pierre Rainville was able to separate the neural correlates of pain-related feelings—"Pain affect," defined as the unpleasantness of pain, the desire to terminate it—from the plain sensation of pain. "Pain affect" engaged the cingulate cortex and insula, while "pain sensation" engaged mostly the SI cortex, a region that we believe to be less importantly involved in emotion. Pierre Rainville, Gary H. Duncan, Donald D. Price, Benoît Carrier, M. Catherine Bushnell, "Pain affect encoded in human anterior cingulate but not somatosensory cortex," *Science* 277 (1997): 968–71.

14 Derek Denton, Robert Shade, Frank Zamarippa, Gary Egan, John Blair-West, Michael McKinley, Jack Lancaster, Peter Fox, "Neuroimaging of genesis and satiation of thirst and an interoceptor-driven theory of origins of primary consciousness," *Proceedings of the National Academy of Sciences* 96 (1999): 5304–9.

15 Terrence V. Sewards, Mark A. Sewards, "The awareness of thirst: proposed neural correlates," *Consciousness & Cognition: An International Journal* 9 (2000): 463–87.

16 Balwinder S. Athwal, Karen J. Berkley, Imran Hussain, Angela Brennan, Michael Craggs, Ryuji Sakakibara, Richard S. J. Frackowiak, Clare J. Fowler, "Brain responses to changes in bladder volume and urge to void in healthy men," *Brain* 124 (2001): 369–77; Bertil Blok, Antoon T. M. Willemsen, Gert Holstege, "A PET study on brain control of micturition in humans," *Brain* 120 (1997): 111–21.

17 Sherif Karama, et al., "Areas of brain activation in males and females during viewing of erotic film excerpts" (cited earlier).

18 David H. Hubel, *Eye, Brain and Vision* (New York: Scientific American Library, 1988).

19   John S. Morris provides a brief review of the current state of affairs in *Trends in Cognitive Sciences* 6 (2002): 317–19.

20   A. D. Craig has proposed that the pathways into the insula use a dedicated thalamic nucleus, VMpo, to project to the insular cortex. Within the insular cortex, the signals brought by these pathways are processed in successive subregions from the back to the front of this sector. This resembles the subregional organization of visual pathways in the occipital cortex beyond the primary visual cortex (VI). In other words, feelings probably depend on processing in interconnected subregions much like vision does.

21   Arthur D. Craig, "How do you feel? Interoception: the sense of the physiological condition of the body," *Nature Reviews* 3 (2002): 655–66; D. Andrew, Arthur D. Craig, "Spinothalamic lamina I neurons selectively sensitive to histamine: a central neural pathway for itch," *Nature Neuroscience* 4 (2001): 72–77; Arthur D. Craig, Kewei Chen, Daniel J. Bandy, Eric M. Reiman, "Thermosensory activation of insular cortex," *Nature Neuroscience* 3 (2000): 184–90.

22   Alain Berthoz, *Le Sens du Mouvement* (Paris: Editions Odile Jacob, 1997).

23   Antoine Lutz, Jean-Philippe Lachaux, Jacques Martinerie, Francisco Varela, "Guiding the study of brain dynamics by using first-person data: synchrony patterns correlate with ongoing conscious states during a simple visual task," *Proceedings of the National Academy of Science* 99 (2002): 1586–91.

24   Richard Bandler, Michael T. Shipley, "Columnar organization in the rat midbrain periaqueductal gray: modules for emotional expression?" *Trends in Neurosciences* 17 (1994): 379–89; Michael M. Behbehani, "Functional characteristics of the midbrain periaqueductal gray," *Progress in Neurobiology* 46 (1995): 575–605.

25   Vittorio Gallese, "The shared manifold hypothesis," *Journal of Consciousness Studies*, 8 (2001): 33–50. Giacomo Rizzolatti, Luciano Fadiga, Leonardo Fogassi, Vittorio Gallese, "Resonance behaviors and mirror neurons," *Archives Italiennes de Biologie* 137 (1999): 85–100; Giacomo Rizzolatti, Leonardo Fogassi, Vittorio Gallese, "Neurophysiological mechanisms underlying the understanding and imitation of action," *Nature Reviews Neuroscience* 2 (2001): 661–70; Giacomo Rizzolatti, Luciano Fadiga, Vittorio Gallese, Leonardo Fogassi, "Premotor cortex and the recognition of motor actions,"

Cognitive Brain Research 3 (1996): 131–41; Ritta Haari, Nina Forss, Sari Avikainen, Erika Kirveskari, Stephan Salenius, Giacomo Rizzolatti, "Activation of human primary motor cortex during action observation: a neuromagnetic study," *Proceedings of the National Academy of Sciences* 95 (1998): 15061–65.

26 Ralph Adolphs, et al. (cited earlier).

27 See Antonio Damasio, *Descartes' Error: Emotion, Reason, and the Human Brain* (cited earlier), *The Feeling of What Happens: Body, Emotion, and the Making of Consciousness* (cited earlier).

28 Ulf Dimberg, Monika Thunberg, Kurt Elmehed, "Unconscious facial reactions to emotional facial expressions," *Psychological Science* 11 (2000): 86–89.

29 Taco J. DeVries, Toni S. Shippenberg, "Neural systems underlying opiate addiction," *Journal of Neuroscience* 22 (2002): 3321–25; Jon-Kar Zubieta, Yolanda R. Smith, Joshua A. Bueller, Yanjun Xu, Michael R. Kilbourn, Douglas M. Jewett, Charles R. Meyer, Robert A. Koeppe, Christian S. Stohler, "Regional mu opioid receptor regulation of sensory and affective dimensions of pain," *Science* 293 (2001): 311–15; Jon-Kar Zubieta, Yolanda R. Smith, Joshua A. Bueller, Yanjun Xu, Michael R. Kilbourn, Douglas M. Jewett, Charles R. Meyer, Robert A. Koeppe, Christian S. Stohler, "Mu-opioid receptor-mediated antinociception differs in men and women," *Journal of Neuroscience* 22 (2002): 5100–7.

30 Wolfram Schultz, Léon Tremblay, Jeffrey R. Hollerman, "Reward prediction in primate basal ganglia and frontal cortex," *Neuropharmacology* 37 (1998): 421–29; Ann E. Kelley and Kent C. Berridge, "The neuroscience of natural rewards: Relevance to addictive drugs," *Journal of Neuroscience* 22 (2002): 3306–11.

31 These responses are quite comparable across subjects. A number of Web sites concerned with drug addiction provide descriptions of drug experiences: http://www.erowid.org/index.shtml.

32 DeVries and Shippenberg, ibid.

33 The activiation of the insula probably is the key correlate of feeling. The cingulate activation probably correlates largely with the regulatory response engaged by the drugs. Naturally, the respiratory responses become part of what is felt. Alex Gamma, Alfred Buck, Thomas Berthold, Daniel Hell, Franz X. Vollenweider, "3,4-

methylenedioxymethamphetamine (MDMA) modulates cortical and limbic brain activity as measured by $[H_2{}^{15}O]$-PET in healthy humans," *Neuropsychopharmacology* 23 (2000): 388–95; Louise A. Sell, John S. Morris, Jenny Bearn, Richard J. Frackowiak, Karl J. Friston, Raymond J. Dolan, "Neural responses associated with cue-invoked emotional states and heroin in opiate addicts," *Drug and Alcohol Dependence* 60 (2000): 207–16; Bruce Wexler, C. H. Gottschalk, Robert K. Fulbright, Isak Prohovnik, Cheryl M. Lacadie, Bruce J. Rounsaville, John C. Gore, "Functional magnetic resonance imaging of cocaine craving," *American Journal of Psychiatry* 158 (2001): 86–95; Luis C. Maas, Scott E. Lukas, Marc J. Kaufman, Roger D. Weiss, Sarah L. Daniels, Veronica W. Rogers, Thellea J. Kukes, Perry F. Renshaw, "Functional magnetic resonance imaging of human brain activation during cue-induced cocaine craving," *American Journal of Psychiatry* 155 (1998): 124–26; Anna Rose Childress, P. David Mozley, William McElgin, Josh Fitzgerald, Martin Reivich, Charles P. O'Brien, "Limbic activation during cue-induced cocaine craving," *American Journal of Psychiatry* 156 (1999): 11–18; Daniel S. O'Leary, Robert I. Block, Julie A. Koeppel, Michael Flaum, Susan K. Schultz, Nancy C. Andreasen, Laura Boles Ponto, G. Leonard Watkins, Richard R. Hurtig, Richard D. Hichwa, "Effects of smoking marijuana on brain perfusion and cognition," *Neuropsychopharmacology* 26 (2002): 802–16.

34   Gerald Edelman (cited earlier), and Rodney A. Brooks, *Flesh and Machines* (New York: Pantheon Books, 2002).

CHAPTER 4: Ever Since Feelings

1   The term *laetitia* is reasonably translated as joy or elation (the latter is Amélie Rorty's proposed translation in *Spinoza on the Pathos of Idolatrous Love and the Hilarity of True Love,* in Amélie Rorty, ed., *Explaining Emotions* (Berkeley: University of California Press, 1980). *Laetitia* also has been translated as pleasure, which is incorrect in my view. *Tristitia* is best translated as sadness or sorrow, although it may point more generally to negative affects such as fear and anger.

When Spinoza mentions greater or lesser perfection he tends to add the word "transition." This is a valuable qualification that draws attention to the dynamic nature of the affect process, but it can be misleading, as if the transitions themselves were the most important part of the processes.

2   It is interesting to note that in the modern field of neural networks certain states of operation have been described as "harmonious." There are even "maximal harmonious states" to be found. The essence of the harmony is the same in biological and artificial operations: ease, efficiency, rapidity, power.

3   For a description of depression as sickness behavior see Bruce G. Charlton, "The malaise theory of depression: major depressive disorder is sickness behavior and antidepressants are analgesic," *Medical Hypotheses* 54 (2000): 126–30. For descriptions of the experience of depression see William Styron, *Darkness Visible: A Memoir of Madness* (New York: Random House, 1990); Kay Jamieson, *An Unquiet Mind* (New York: Knopf, 1995); and Andrew Solomon, *The Noonday Demon: An Anatomy of Depression* (London: Chatto & Windus, 2001).

4   See Antonio Damasio, *Descartes' Error: Emotion, Reason, and the Human Brain* (cited earlier); Antonio Damasio, "The somatic marker hypothesis and the possible functions of the prefrontal cortex" (cited earlier).

5   Antoine Bechara et al., "Insensitivity to future consequences following damage to human prefrontal cortex" (cited earlier); Antonio Damasio, Steven Anderson, "The frontal lobes," in K. M. Heilman and E. Valenstein (eds.), *Clinical Neuropsychology, Fourth Edition* (New York: Oxford University Press, 2002); Facundo Manes, Barbara Sahakian, Luke Clark, Robert Rogers, Nagui Antoun, Mike Aitken, Trevor Robbins, "Decision-making processes following damage to the prefrontal cortex," *Brain* 125 (2002): 624–39; Daniel Tranel, Antoine Bechara, Natalie Denburg, "Asymmetric functional roles of right and left ventromedial prefrontal cortices in social conduct, decision-making, and emotional processing," *Cortex* (in press).

6   For details on the neural and cognitive aspects of working memory, see Patricia Goldman-Rakic, "Regional and cellular fractionation of

working memory," *Proceedings of the National Academy of Sciences of the United States of America* 93 (1996): 13473–80; and Alan Baddeley, "Recent developments in working memory," *Current Opinion in Neurobiology* 8 (1998): 234–38. For a general treatment of prefrontal cortex functions see Joaquin Fuster, *Memory in the Cerebral Cortex* (Cambridge, Mass., London, UK: MIT Press, 1995); and Elkhonon Goldberg, *The Executive Brain: Frontal Lobes and the Civilized Mind* (New York: Oxford University Press, 2001).

7   Jeffrey Saver, Antonio Damasio, "Preserved access and processing of social knowledge in a patient with acquired sociopathy due to ventromedial frontal damage," *Neuropsychologia* 29 (1991): 1241–49.

8   Antonio Damasio, *Descartes' Error: Emotion, Reason, and the Human Brain* (cited earlier).

9   When I began presenting these ideas, about two decades ago, they were greeted with a mixture of puzzlement and resistance. At first the evidence was anecdotal and I had no support from the previous literature with the exception of an article by the neuroanatomist Walle Nauta on the possible role of the frontal cortex in emotion, "The problem of the frontal lobe: a reinterpretation," *Journal of Psychiatric Research* 8 (1971): 167–87. The evidence has now grown and so has the endorsement of these ideas, e.g., Antoine Bechara, et al., "Insensitivity to future consequences following damage to human prefrontal cortex" (cited earlier); Antoine Bechara, et al., "Failure to respond autonomically to anticipated future outcomes following damage to prefrontal cortex" (cited earlier); Antoine Bechara, et al., "Deciding advantageously before knowing the advantageous strategy" (cited earlier); Antoine Bechara, Hanna Damasio, Antonio R. Damasio, Greg P. Lee, "Different contributions of the human amygdala and ventromedial prefrontal cortex to decision-making," *Journal of Neuroscience* 19 (1999): 5473–81; Antoine Bechara, Hanna Damasio, Antonio Damasio, "Emotion, decision-making, and the orbitofrontal cortex," *Cerebral Cortex* 10 (2000): 295–307; Shibley Rahman, Barbara J. Sahakian, Rudolph N. Cardinal, Robert D. Rogers, Trevor W. Robbins, "Decision making and neuropsychiatry," *Trends in Cognitive Sciences* 5 (2001): 271–77; Geir Overskeid, "The slave of the passions: experiencing problems and selecting solutions," *Review of General Psychology* 4 (2000): 284–309; George Loewenstein, E. U.

Webber, C. K. Hsee, "Risk as feelings," *Psychological Bulletin* 127 (2001): 267–86; Jean-P. Royet, David Zald, Rémy Versace, Nicolas Costes, Frank Lavenne, Olivier Koenig, Rémi Gervais, "Emotional responses to pleasant and unpleasant olfactory, visual, and auditory stimuli: a positron emission tomography study," *Journal of Neuroscience* 20 (2000): 7752–59.

10   Stefan P. Heck, *Reasonable Behavior: Making the Public Sensible* (University of California, San Diego, 1998); Ronald de Sousa, *The Rationality of Emotion* (Cambridge: MIT Press, 1991); Martha Nussbaum, *Upheavals of Thought* (cited earlier).

11   Ralph Adolphs, et al., "Impaired recognition of emotion in facial expressions following bilateral damage to the human amygdala" (cited earlier).

12   James K. Rilling, David A. Gutman, Thorsten R. Zeh, Giuseppe Pagnoni, Gregory S. Berns, Clinton D. Kilts, "A neural basis for social cooperation," *Neuron* 35 (2002): 395–405.

13   Steven Anderson, Antoine Bechara, Hanna Damasio, Daniel Tranel, Antonio Damasio, "Impairment of social and moral behavior related to early damage in human prefrontal cortex," *Nature Neuroscience* 2 (1999): 1032–37.

14   This interpretation is reinforced by evidence from other patients who sustained damage to the brain regions that inform the prefrontal cortices of the premises of a situation, e.g., the right inferotemporal sector. In collaboration with my colleagues Steven Anderson and Hanna Damasio, I have noted that damage to this sector during the developmental period arrests the maturation of proper social behavior. The practical result can be comparable to that of adult prefrontal lesions.

15   Jonathan Haidt, "The Moral Emotions," in R. J. Davidson, K. Scherer, and H. H. Goldsmith (eds.), *Handbook of Affective Sciences* (Oxford University Press, in press); R. A. Shweder and J. Haidt, "The cultural psychology of the emotions: Ancient and new," in M. Lewis & J. Haviland (eds.), *Handbook of emotions*, 2nd Ed. (New York: Guilford, 2000).

16   E. O. Wilson's "consilience" project is an example of the sort of attitude that could advance knowledge by bringing together biology and the humanities. Edward O. Wilson, *Consilience* (New York: Knopf, 1998).

17 I note that all of these comments on ethics apply to ethical behaviors
and to their possible biological origins and mechanisms within the
rubric of descriptive ethics. I am not referring here to issues of nor-
mative ethics or metaethics.

18 Frans de Waal, *Good Natured* (cited earlier); B. Heinrich, *The Mind
of the Raven* (cited earlier); Hans Kummer, *The Quest of the Sacred
Baboon* (cited earlier). The altruism experiment in rhesus monkeys
is discussed by Marc Hauser in *Wild Minds* (New York: Holt and
Company, 2000) and was conducted by Robert Miller (R. E. Miller,
J. Banks, H. Kuwhara, "The communication of affect in monkeys:
Cooperative conditioning," *Journal of Genetic Psychology* 108 [1966]:
121–34; R. E. Miller, "Experimental approaches to the physiological
and behavioral concomitants of affective communication in rhesus
monkeys," in S. A. Altmann [ed.], *Social Communication among Primates* [Chicago: University of Chicago Press, 1967]).

19 Genes are not just necessary to construct a certain kind of brain
equipped with the devices we have been discussing. The expression
of genes is also necessary to permit learning and the renewal and repair of brain structure. In addition, gene expression is dependent on
interactions with the environment throughout development and ma-
turity. A comprehensive view of the issues discussed here benefits
from an extensive, diverse, and at times polemic literature in the
fields of evolutionary psychology, neurobiology, and population genetics. Important readings listed in chronological order include:
William Hamilton, "The genetical evolution of social behaviour,"
Parts 1 and 2, *Journal of Theoretical Biology* 7 (1964): 1–52; George
Williams, *Adaptation and Natural Selection: A Critique of Some Current
Evolutionary Thought* (Princeton, N.J.: Princeton University Press,
1966); Edward O. Wilson, *Sociobiology: The New Synthesis* (Cam-
bridge, Mass.: Harvard University Press, 1975); Richard Dawkins,
The Selfish Gene (New York: Oxford University Press, 1976); Stephen
Jay Gould, *The Mismeasure of Man* (New York: Norton, 1981); Steven
Rose, Richard Lewontin, Leo Kamin, *Not in Our Genes* (Harmonds-
worth: Penguin, 1984); Leda Cosmides, John Tooby, *The Adapted
Mind: Evolutionary Psychology and the Generation of Culture* (New York:
Oxford University Press, 1992); Helena Cronin, John Smith, *The Ant
and the Peacock: Altruism and Sexual Selection from Darwin to Today*
(Cambridge, U.K.: Cambridge University Press, 1993); Richard C.

Lewontin, *Biology as Ideology: The Doctrine of DNA* (New York: Harper-Collins, 1992); Carol Tavris, *The Mismeasure of Women* (New York: Simon and Schuster, 1992); Robert Wright, *The Moral Animal: Why We Are the Way We Are: The New Science of Evolutionary Psychology* (New York: Pantheon Books, 1994); Mark Ridley, *Evolution* (Oxford, England; New York: Oxford University Press, 1997); Steven Rose, *Lifelines: Biology, Freedom, Determinism* (Harmondsworth: Allen Lane, 1997); Edward O. Wilson, *Consilience* (cited earlier); Steven Pinker, *How the Mind Works* (New York: W. W. Norton & Company, 1998); Patrick Bateson and Martin Paul, *Design for a Life: How Behaviour Develops* (London: Jonathan Cape, 1999); Hilary Rose and Steven Rose, eds., *Alas, Poor Darwin* (New York: Harmony Books, 2000); Melvin Konner, *The Tangled Wing* (New York: Henry Holt and Company, 2002); Robert Trivers, *Natural Selection and Social Theory: Selected Papers of Robert L. Trivers* (New York: Oxford University Press, 2002).

20 See Martha Nussbaum, *Upheavals of Thought* (Cambridge University Press, 2001) for a discussion of the role of emotions in justice, in general, and in the application of justice, in particular.

21 William Safire has recently used the term "neuroethics" to refer to the debate for the ethical issues raised by new therapies of neurological and psychiatric disorders. That debate will be informed by some of the issues discussed here but the goals of "neuroethics" and my discussion are different. More than a decade ago Jean Pierre Changeux used the term neuroethics to denote the matter discussed in this chapter at the time of a landmark symposium on biology and ethics held in Paris under the auspices of the Institute Pasteur.

22 The blossoming of the new means of social governance probably was ushered in by phenomena as disparate as climate changes and the developments of symbolization and agriculture. For a treatment of those important factors see William Calvin, *The Ascent of Mind: Ice Age Climates and the Evolution of Intelligence* (New York: Bantam Books, 1991), *A Brain for All Seasons: Human Evolution and Abrupt Climate Change* (Chicago; London: University of Chicago Press, 2002); Terrence Deacon, *The Symbolic Species: The Co-evolution of Language and the Brain* (New York: W. W. Norton & Company, 1997); and Jared Diamond, *Guns, Germs, and Steel: The Fates of Human Societies* (New York: W. W. Norton & Company, 1997).

23 Although a discussion of the historical connection of these ideas is
    beyond my expertise, I should note a bridge to two among the es-
    tablished views regarding ethics and the related apparatus of justice:
    the Scottish Enlightenment view, and the Kantian view. According
    to the Scottish Enlightenment view, justice is grounded in emotion,
    specifically in positive moral emotions such as sympathy that are
    part and parcel of natural human behavior. One can cultivate moral
    emotions, but they do not need to be taught. They are largely innate,
    part of the evolved natural goodness of humanity. On the basis of
    such emotions and with the obvious help of knowledge and reason,
    rules of ethics, laws, and systems of justice are eventually codified.
    Adam Smith and David Hume are the prominent exponents of this
    view, although it is apparent that the beginnings of the idea can be
    found in Aristotle. (Adam Smith, *A Theory of Moral Sentiment*
    [Cambridge, U.K.; New York: Cambridge University Press, 2002];
    David Hume, *A Treatise of Human Nature; Enquiry Concerning the
    Principles of Morals* [Garden City, N.Y.: Doubleday, 1961]; Aristotle,
    Nichomachean Ethics.)

    The other view is identified with Kant and its modern expres-
    sion can be found in the work of John Rawls. It rejects emotions as
    a possible foundation for justice, choosing reason instead as the
    only proper grounding for ethics, laws, and justice. The Kantian
    view does not trust emotion of any kind, deems it capricious, even
    dangerous. Kant rejects the wisdom of emotions, the fine and pa-
    tient job with which evolution has amassed some useful guidelines
    for the governance of social life. It should be said, however, that
    Kant also rejects the not-so-wise and the cruel aspects of nature as
    expressed in the apparatus of emotion. His sweeping rejection guar-
    antees that he will not be fooled by natural moral emotions. Instead,
    he trusts human reason and creativity to invent better solutions
    than evolution alone ever did, or perhaps ever could, without delib-
    erate human effort. Therein lies the problem, because unfeelingly
    tempered reason can be just as bad a counselor as natural emotions.
    See Robert Wright, *The Moral Animal: Why We Are the Way We Are:
    The New Science of Evolutionary Psychology* (New York: Random
    House, 1994), for an incisive discussion of the perils of trusting all
    that comes naturally in the realm of ethics. See Jonathan Haidt, for

a review of Kantian and Humean perspectives on moral judgment. "The Emotional Dog and Its Rational Tail," *Psychological Review* 198 (2001): 814–34. See also Paul M. Churchland, *Rules, Know-How, and the Future of Moral Cognition* in *Moral Epistemology Naturalized*, ed. Richmond Campbell and Bruce Hunter (Calgary: University of Calgary Press, 2000); Robert C. Solomon, *A Passion for Justice* (Boston: Addison-Wesley, 1990); John Rawls, *A Theory of Justice* (Cambridge, Mass.: Belknap Press of Harvard University Press, 1971).

The Scottish view has limitations as well. The picture painted by the Scottish view is a bit too optimistic. It uses less of the nasty and brutish conception of humanity that Thomas Hobbes emphasized than of the goodness and nobility of humans that we associate with Jean-Jacques Rousseau, although it cannot be confused with the latter. Beyond the "positive" moral emotions that the Scots emphasize, there are "negative" moral emotions as well, for example, resentment, revenge, and indignation, which are just as relevant to the construction of justice. I see the role of emotions and feelings in justice as going well beyond the evolutionarily inherited moral emotions. I see primary sorrow and joy as having played, and still playing, a principal role in the construction of justice. The personal experience of sorrow in relation to loss, for example, allows us to comprehend the sorrow of the other. Natural sympathy tunes us into the problem of the other, but personally felt pain deepens our sense of the pain expressed and felt by someone else. In other words, personal sorrow would allow us to move from sympathy to empathy. Personal sorrow also would be a most effective springboard for reasoning about the circumstances that cause sorrow and about the means to prevent it in the future. The information provided by emotions and feelings can be used not only to create better instruments of justice, but to create conditions in which justice is more easily possible.

24 Spinoza, *A theologico-political treatise*, 1670. From the R. H. M. Elwes translation, *Benedict de Spinoza: A theologico-political treatise and a Political Treatise* (cited earlier).

25 James L. McGaugh, Larry Cahill, Benno Roozendaal, "Involvement of the amygdala in memory storage: interaction with other brain

systems," (review) *Proceedings of the National Academy of Sciences of the United States of America* 93 (1996): 13508–14. James L. Mc-Gaugh, Larry Cahill, Benno Roozendaal, "Involvement of the amygdala in memory storage: interaction with other brain systems," *Proceedings of the National Academy of Sciences of the United States of America* 93 (1996): 13508–14; Ralph Adolphs, Larry Cahill, Rina Schul, Ralf Babinski, "Impaired memory for emotional stimuli following bilateral damage to the human amygdala," *Learning and Memory* 4 (1997): 291–300; Kevin S. LaBar, Joseph E. LeDoux, Dennis D. Spencer, Elizabeth A. Phelps, "Impaired fear conditioning following unilateral temporal lobectomy in humans," *Journal of Neuroscience* 15 (1995): 6846–55; Antoine Bechara, Daniel Tranel, Hanna Damasio, Ralph Adolphs, Charles Rockland, Antonio Damasio, "A double dissociation of conditioning and declarative knowledge relative to the amygdala and hippocampus in humans," *Science* 269 (1995): 1115–18.

CHAPTER 5: Body, Brain, and Mind

1   In *The Feeling of What Happens: Body, Emotion, and the Making of Consciousness* I elaborate on the mind/consciousness distinction (Antonio Damasio, 2000, cited earlier). I also introduce the notions of core self and extended, or autobiographical, self.

2   The mind-body problem has been considered in great detail by contemporary philosophers of mind, among them David Armstrong, *The Mind-Body Problem: An Opinionated Introduction* (Oxford, U.K., Boulder, Colorado: Westview Press, 1999); Paul Churchland and Patricia Churchland, *On the Contrary* (Boston: MIT Press, 1998); Patricia Churchland, *Brain-Wise* (Cambridge, Mass.: MIT Press, 2002); Patricia Churchland, Paul Churchland, "Neural worlds and real worlds" (*Nature Neuroscience Reviews*, 2002); Daniel Dennett, *Consciousness Explained* (Boston: Little Brown, 1991); David Chalmers, *The Conscious Mind* (New York: Oxford University Press, 1996); Thomas Metzinger, *Conscious Experience* (Paderborn, Germany: Imprint Academic/Schoeningh, 1995); Colin McGinn, *The Problem of Consciousness* (New York: Oxford University Press, 1991);

Galen Strawson, *Mental Reality* (Cambridge, Mass.: MIT Press, 1994); Ned Block, Owen Flanagan, Güven Güzeldere, eds., *The Nature of Consciousness: Philosophical Debates* (Cambridge, Mass.: MIT Press, 1997); and John Searle, *The Rediscovery of the Mind* (Boston: MIT Press, 1992); by philosophers of the recent past: Herbert Feigl, *The 'Mental' and the 'Physical'* (Minneapolis: University of Minnesota Press, 1958); Edmund Husserl, *The Phenomenology of Internal Time-consciousness* (Bloomington, Ind.: Indiana University Press, 1964); Maurice Merleau-Ponty, *Phenomenology of Perception*, trans. by Colin Smith (London: Routledge and Kegan Paul, 1962); and by modern biologists, among them, Jean Piaget, *Biology and Knowledge: An Essay on the Relations between Organic Regulations and Cognitive Processes* (Chicago: University of Chicago Press, 1971); Jean Pierre Changeux, *Neuronal Man: The Biology of Mind* (New York: Pantheon, 1985); Francis Crick, *The Astonishing Hypothesis: The Search for the Soul* (New York: Scribner, 1994); Gerald Edelman, *Remembered Present* (New York: Basic Books, 1989), and *Bright Air, Brilliant Fire: On the Matter of the Mind* (New York: Basic Books, 1992); Francisco Varela, "Neurophenomenology: A methodological remedy to the hard problem," *Journal of Consciousness Studies* 3 (1996): 330–50; Francisco Varela and Jonathan Shear, "First-person methodologies: why, when and how," *Journal of Consciousness Studies* 6 (1999): 1–14.

3   The New Church was one of the first Protestant churches built in Holland (1649–56), and it was truly new, designed from the ground up as a celebration of the Reformed Church. It was not a Catholic Church stripped of its decorations. Today it has become one of the main venues for cultural events in The Hague. The conflict behind the architecture is apparent, and is typical of this age. In accordance with the Reformed Church aesthetic, the building was to be a rejection of ostentation; but as an affirmation of that same Church, the building could hardly have been modest. A similar conflict is apparent thirty miles to the northeast, in the Portuguese synagogue of Amsterdam, another building of the same age (completed in 1675), and just as torn between modesty and pride. The predictable result is that the New Church is both bare and imposing. From the raised altar, which is used as a stage, the podium commands a sweeping view of the large space.

4 René Descartes, Correspondence with Princess Elizabeth of Bohemia. *Oeuvres et lettres* (Bruges, Librarie Gallimard, 1952), and *Meditations and Other Metaphysical Writings* (London: Penguin Books, 1998).

5 Gilbert Keith Chesterton, *The Innocence of Father Brown* (New York: Dodd, Mead, 1911).

6 The neurosurgeon Wilder Penfield studied this phenomenon in several epileptic patients he was attempting to treat. The process probably begins in the cortex of the insula and eventually takes over other sectors of the somatosensing complex, an idea that is compatible with the new findings discussed in Chapter Three. Wilder Penfield, Herbert Jasper, *Epilepsy and the Functional Anatomy of the Human Brain* (Boston: Little, Brown, 1954).

7 The alternative interpretation is that the loss of consciousness is unrelated to the changes in body sensation, and that it would have occurred even if the changed body sensation had not. Loss of consciousness does occur in varied kinds of seizure without any body aura. That is compatible, however, with the notion that, in this kind of seizure, loss of consciousness occurs because body input is inactivated, in advance of other seizure mechanisms causing other manifestations such as convulsions.

8 Oliver Sacks, in *A Leg to Stand On* (London: Duckworth, 1984), and Vilayanur Ramachandran, in *Phantoms in the Brain* (New York: HarperCollins, 1999), have described alterations of limb perception in detail.

9 Sacks's patient suffered a loss of her proprioceptive sense caused by involvement of nerve pathways signaling from her muscles into the central nervous system. Oliver Sacks, in *The Man Who Mistook His Wife for a Hat* (New York: Summit Books, 1985). There also is interesting new evidence that so-called out-of-body experiences can be triggered by direct electrical stimulation of the right somatosensory cortices, namely in the angular gyrus territory. A patient so stimulated reported a separation between the experience of her own body and other mental activities. During stimulation, she imagined herself transposed to the ceiling of her bedroom, from where she was able to observe part of her own body. These findings add to the notion that our sense of the body depends on neural mappings within a dedicated, multicomponent system. Parts of the system are located

in the right cerebral cortex, other parts are located in subcortical regions. Dysfunction involving most of the system, at cortical level, breaks down the sense of our own body and disrupts mind processes. Dysfunction restricted to a sector results in partial syndromes such as asomatognosia and in odd experiences such as out-of-body states. Extensive subcortical dysfunction, as in cases of extensive damage to the brain stem tegmentum, tends to disrupt the system most extensively. See Olaf Blanke et al., "Leaving your body behind," *Nature* (2002), in press.

10   Maps and representations. Antonio Damasio, Hanna Damasio, "Cortical systems for retrieval of concrete knowledge: the convergence zone framework," *Large-Scale Neuronal Theories of the Brain*, Christof Koch (ed.) (Cambridge: MIT Press, 1994): 61–74; Antonio Damasio, "Time-locked multiregional retroactivation: A systems level proposal for the neural substrates of recall and recognition," *Cognition* 33 (1989): 25–62; Antonio Damasio, "The brain binds entities and events by multiregional activation from convergence zones," *Neural Computation* 1 (1989): 123–32.

11   See Francis Crick, *The Astonishing Hypothesis: The Search for the Soul* (cited earlier); Giulio Tononi and Gerald Edelman, "Consciousness and complexity," *Science* 282 (1998): 1846–51; and Jean Pierre Changeux, Paul Ricoeur, *Ce qui nous fait penser, La nature et la règle* (Paris: Editions Odile Jacob, 1998), for treatments of this issue. See Antonio Damasio, *The Feeling of What Happens: Body, Emotion, and the Making of Consciousness* (cited earlier) for a discussion of the problems facing the neurobiological investigation of consciousness.

12   The notion that both the processes of learning and those of perception are based on "selections" of neuronal elements of a preexisting repertoire is relatively recent. See Jean-Pierre Changeux, *Neuronal Man: The Biology of Mind* (cited earlier); Gerald Edelman, *Neural Darwinism: The Theory of Neuronal Group Selection* (New York: Basic Books, 1987).

13   David Hubel, *Eye, Brain and Vision* (cited earlier).

14   Roger B. Tootell, Eugene Switkes, Michael S. Silverman, Susan L. Hamilton, "Functional anatomy of macaque striate cortex. II. Retinotopic organization," *The Journal of Neuroscience* 8 (1988): 1531–68.

15   Joanna Aizenberg, Alexei Tkachenko, Steve Weiner, Lia Addadi, Gordon Hendler, "Calcitic microlenses as part of the photoreceptor system in brittlestars," *Nature* 412 (2001): 819–22; Roy Sambles, "Armed for light sensing," *Nature* 412 (2001): 783.

16   Samer Hattar, Hsi-Wen Liao, Motoharu Takao, David M. Berson, King-Wai Yau, "Melanopsin-containing retinal ganglion cells: architecture projections, and intrinsic photosensitivity," *Science* 295 (2002): 1065–70; David M. Berson, Felice Dunn, Motoharu Takao, "Phototransduction by retinal ganglion cells that set the circadian clock," *Science* 295 (2002): 1070–73.

17   Nicholas Humphrey, *A History of the Mind* (New York: Simon and Schuster, 1992).

18   David Hubel, Margaret Livingstone, "Segregation of form, color, and stereopsis in primate area 18," *The Journal of Neuroscience* 7 (1987): 3378–415; Semir Zeki (*Vision of the Brain*); R. Wurtz; R. Desimone.

19   George Lakoff, Mark Johnson, *Metaphors We Live By* (Chicago: University of Chicago Press, 1980), and George Lakoff, Mark Johnson, *Philosophy in the Flesh* (New York: Basic Books, 1999); Mark Johnson, *The Body in the Mind* (Chicago: University of Chicago Press, 1987).

20   Hubel, ibid.

21   This view also requires a qualification regarding the type of reductionism we are using in this exercise. The mind level of biological phenomena has additional specifications that are not present at the neural-map level. I hope a reductionist research strategy eventually will allow us to explain how we get from the "neural-map" level to the "mental" level, although the mental level will not "reduce to" the neural-map level because it possesses emergent properties created *from* the neural-map level. There is nothing magic about those emergent properties, but there is a lot that remains mysterious, given our massive ignorance of what they may involve.

22   Spinoza, *The Ethics* (cited earlier).

23   For a presentation of this idea and for a discussion of its possible neural implementation, see Antonio Damasio, *The Feeling of What Happens: Body, Emotion, and the Making of Consciousness* (cited earlier).

24   Spinoza, *The Ethics*, Part III (cited earlier).

25   In *Behind the Geometric Method: A Reading of Spinoza's Ethics* (cited earlier), Edwin Curley provides a reading of Spinoza's thinking that

would be compatible with this view. So does Gilles Deleuze in *Spinoza: A Practical Philosophy* (cited earlier).

26 The immortality of the mind plays a curious and uneven role in the history of Jewish thinking. In Spinoza's time, denying the immortality of the mind was indeed a heresy, for rabbis and lay leaders of the community, and created a problem for the Christian community that had welcomed the Jews in Holland. See Steven Nadler, *Spinoza's Heresy* (New York: Oxford University Press, 2002), for an illuminating treatment of this issue.

27 Simon Schama, *Rembrandt's Eyes* (New York: Knopf, 1999).

28 For a very different and fascinating interpretation of what is going on in this painting see W. G. Sebald's *The Rings of Saturn*. Sebald believes Rembrandt deliberately undermined Tulp and his colleagues—caught desecrating a body—by lovingly lighting the face of Aris Kindt, the unfortunate thief who had been hanged just a few hours before and who was not participating in the proceedings of his own free will. But Sebald is not correct about his claim that Rembrandt made a deliberate error in the depiction of Kindt's left hand, which is entirely correct. Winfried Georg Sebald, *The Rings of Saturn* (New York: New Directions Publishing Corporation, 1998).

CHAPTER 6: A Visit to Spinoza

1 Albert Einstein, *The World as I See It* (New York: Covici Friede Publishers, 1934).

2 Alfred North Whitehead, *Science and the Modern World* (New York: Macmillan, 1967).

3 Diogo Aurélio argues the point convincingly (*Imaginação e Poder*, Lisbon: Colibri, 2000). See also Carl Gebhardt, "Rembrandt y Spinoza," *Revista de Occidente*.

4 Simon Schama, *An Embarrassment of Riches* (cited earlier).

5 Hana Debora was the second wife of Miguel de Espinoza and was half his age. She descended from an impressive line of physicians, philosophers, and theologians and was raised in the northern Portuguese city of Porto by her mother Maria Nunes. She came to Amsterdam to marry Spinoza's father, who had just been widowed, and bear his children.

6   In *Um Bicho da Terra* (Lisbon: Guimãres Editores, 1984) Agustina Bessa Luís provides a fictionalized account of life in sixteenth-century Porto that inspired this sentence.

7   Steven Nadler, *Spinoza: A Life* (Cambridge, U.K.; New York: Cambridge University Press, 1999).

8   Marilena Chaui, *A Nervura do Real* (São Paulo: Companhia das Letras, 1999).

9   A. H. de Oliveira Marques, *History of Portugal, Vol. I* (New York: Columbia University Press, 1972); Francisco Bettencourt, *Historia das Inquisicões Portugal, Espanha e Italia XV-XIX* (São Paulo: Companhia das Letras, 1994); Cecil Roth, *A History of the Marranos* (New York: Meridian Books, 1959).

10 Marques, ibid; Bettencourt, ibid; Roth, ibid.

11 Bettencourt, ibid; Antonio José Saraiva; Marques, ibid.

12 Léon Poliakov, *Histoire de l'Antisémistisme,* 3rd ed. (Paris: Calmann-Lévy, 1955).

13 C. Gebhardt, as cited by Gabriel Albiac, *La Synagogue Vide* (Paris: Presses Universitaires de France, 1994).

14 Frederick Pollock, *Spinoza: His Life and Philosophy* (London: C. Kegan Paul & Co., 1880).

15 The least reliable of Spinoza's biographers, Lucas, suggested that Spinoza did compose a response but no trace of it exists. Probably no such response was ever prepared.

16 Luís Machado de Abreu, *A Recepção de Spinoza em Portugal.* In *Sob O Olher de Spinoza* (Aveiro, Portugal: Universidade de Aveiro, 1999).

17 Maria Luisa Ribeiro Ferreira, *A Dinâmica da Razão na Filosofia de Espinosa* (Lisbon: Gulbenkian Foundation, 1997).

18 Jonathan I. Israel, *Radical Enlightenment: Philosophy and the Making of Modernity 1650–1750* (Oxford University Press, 2001).

19 Locke was not a religious radical. He was a believer and offered a safe and noncontentious way of airing some of Spinoza's radical ideas. On the other hand, it is difficult to imagine that Locke was not influenced by Spinoza. He lived his Amsterdam exile from 1683 to 1689, shortly after Spinoza's death, in the period of most heated discussion and scandal regarding Spinoza's ideas. This period preceded the publication of Locke's own work (the *Essays* and the *Two Treatises* only begin to appear in 1690). John Locke, *An Essay Considering Human Understanding* (Oxford: Clarendon Press,

1975); *Two Treatises of Government* (London: Cambridge University Press, 1970).

20   Voltaire, *Les Systèmes,* Oeuvres (Paris: Moland, 1993), 170. The original text reads as follows:

Alors un petit juif, au long nez, au teint blême,

Pauvre, mais satisfait, pensif et retiré,

Esprit subtil et creux, moins lu que célébré

Caché sous le manteau de Descartes, son maître,

Marchant à pas comptés, s'approche du grand être:

Pardonnez-moi, dit-il, en lui parlant tout bas,

Mais je pense, entre nous, que vous n'existez pas.

21   Gabriel Albiac, *La Synagogue Vide* (cited earlier).

22   Johann von Goethe, *The Auto-Biography of Goethe: Truth and Poetry: From My Life,* Parke Godwin, ed. (London: H. G. Bohn, 1848).

23   Georg W. F. Hegel, *Spinoza* translated from the Second German Edition by E. S. Haldane and F. H. Simson (London: Kegan Paul, 1892).

24   Circular of the Spinoza Committee: A Statue to Spinoza. 1876 in Frederick Pollock, *Spinoza: His Life and Philosophy* (London: C. Kegan Paul & Co., 1880), Appendix D.

25   Michael Hagner and Bettina Wahrig-Schmidt, eds., *Johannes Müller und die Philosophie* (Berlin: Akademie Verlag, 1992).

26   Frederick Pollock, ibid.

27   Siegfried Hessing, "Freud et Spinoza," *Revue Philosophique* 2 (1977): 168 (author's translation).

28   Hessing, ibid., 169 (author's translation).

29   Jacques Lacan, *Les Quatre Concepts Fondamentaux de la Psychanalyse* (Paris: Edition Le Seuil, 1973).

30   Albert Einstein, *Out of My Later Years* (New York: Wings Books, 1956).

31   Margaret Gullan-Whur, *Within Reason: A Life of Spinoza* (New York: St. Martin's Press, 2000). Both Stuart Hampshire (*Spinoza,* cited earlier), and Steven Nadler (*Spinoza: A Life,* cited earlier) have suggested that glass dust may have been a factor in his illness.

32   Hampshire, ibid.

33   At no point in this trajectory, in my mind or in that of his leading biographers—Colerus, Pollock, Nadler, Gullan-Whur—can Spinoza

be seen as an autistic person, specifically someone with Asperger syndrome, as has been suggested recently by the psychiatrist Michael Fitzgerald. Michael Fitzgerald, "Was Spinoza Autistic?" *The Philosophers' Magazine*, 14, Spring 2001. Autistic persons have serious social difficulties, tend to lack empathy, and often live a lonely and friendless existence. There is no evidence that Spinoza had social difficulties of any consequence—other than for those that his intellectual achievements created for him *vis-à-vis* his community and the political and ecclesiastical world. He does not appear to have lived in any greater isolation than, say, Descartes when one takes into account his close friendships, his warm inclusion in the Van der Spijk home, and the countless visitors he received daily. There is reason to believe that he was a gregarious young man and a number of passages of his writings also suggest extensive sexual experience in his Amsterdam days. Most importantly, the diagnosis is hardly compatible with Spinoza's deep understanding of the workings of human beings and society. He shows no sign of lack of empathy and even his youthful arrogance and apparent sense of superiority, hardly surprising in a young intellectual in his circumstances, seem to have been curbed with the passing years.

CHAPTER 7: Who's There?

1   The expression suggests that God and Nature are one and the same thing. That is not exactly so, however. Spinoza makes a subtle distinction between the part of nature that is generative and closer to the traditional notion of a creator God—*Natura naturans*—and the part of nature that is the result of creation—*Natura naturata*. See Steven Nadler, *Spinoza's Heresy* (cited earlier), for a discussion of this issue.

2   For Spinoza, salvation occurs personally and privately, but with the help of others in society. And the state can facilitate the personal and social efforts. The state must be democratic, its laws must be fair, and it must allow its citizens to live free from fear. Having politics as subsidiary to the problem of salvation distinguished Spinoza from Hobbes, Spinoza's older contemporary. (See Maria Luisa Ribeiro

Ferreira, *A Dinâmica da Razão na Filosofia de Espinosa* [cited earlier]
for a comment on this distinction.) For Hobbes, a good political system was one that permitted the proper function of a state in which
the individual was a subject. For Spinoza, a good political system
was one that helped a free citizen achieve salvation.

3   Spinoza's correspondence. Letter XLIX in Robert Harvey Monro
Elwes, *Improvement of the Understanding, Ethics and Correspondence
of Benedict de Spinoza* (Washington: Dunne, 1901).

4   Here are Spinoza's words from *A Theologico-political treatise,* 1670
(R. H. M. Elwes translation): "Before I go further I would expressly
state (though I have said it before) that I consider the utility and the
need for Holy Scripture or Revelation to be very great. For as we cannot perceive by the natural light of reason that simple obedience is
the path of salvation, and are taught by revelation only that it is so by
the special grace of God, which our reason cannot attain, it follows
that the Bible has brought a very great consolation to mankind. All
are able to obey, whereas there are but very few, compared with the
aggregate of humanity, who can acquire the habit of virtue under
the unaided guidance of reason. Thus if we had not the testimony of
Scripture, we should doubt of the salvation of nearly all men."

This deeply felt attitude gives the lie to the caricatures of Spinoza as the devil incarnate. Late Spinoza advised those around him,
who were mostly Christian, to remain within their church, largely
the Protestant church. He urged children to attend Mass, and he
himself heard sermons by Colerus, the Lutheran pastor who moved
into the house Spinoza once rented in the Stilleverkade and became
first his friend, then his biographer. Spinoza did not have faith in a
provident God or in eternal life, but he never mocked the faith of
others. In fact, Spinoza was extremely careful with the faith of the
unlearned. He only discussed religion with his intellectual colleagues. As noted, he disallowed Dutch translations of his work in
order to prevent the rapid dissemination of his ideas among those
who might not be prepared to cope with their consequences. In reality, precious few among those that read his Latin originals were
prepared to read him with equanimity, but he certainly tried to
dampen the wildfire impact of his ideas. He refused to be the leading figure of an intellectual movement, which he could have become

had he wished to do so. Would that have pleased him, had it been possible to assume such a public role and remain not just free but alive? In his article on Spinoza, Pierre Bayle (*Dictionnaire Historique et Critique*, Rotterdam, 1702) thought he might have wished to become a public leader. But I doubt this, given the personality of Spinoza I have come to imagine. At least by the time of The Hague, Spinoza would no longer harbor such ambitions.

5 In *Modos de Evidência* (Lisbon: Imprensa Nacional, 1986), Fernando Gil discusses this form of intellectual process and its affective consequences.

Spinoza's solution bears the stamp of many influences. A critical influence is likely to have come from the Greek and Roman Stoic philosophers as Susan James argues convincingly (*The Rise of Modern Philosophy*, Tom Sorrell, ed., Oxford, U.K.: Clarendon Press, 1993). The Judaic influence is patent in the accent placed on life on earth rather than life eternal, in the emphasis given to ethical conduct, and in the tying up of ethical virtue and sociopolitical organization, a consistent trait of Old Testament narratives. Some influence may have come from the Kabbalah. Spinoza was critical of the superstitious aspects of the Kabbalah, but his system does borrow the Kabbalah's reverence for "a mystery without a face," as Maria Luisa Ribeiro Ferreira terms it (*A Dinâmica da Razão na Filosofia de Espinosa*, cited earlier). The Christian influence also is apparent. In Spinoza's system the *amor intellectualis Dei* can only flourish in an individual who behaves on the example of Christ: unconditionally respectful and loving of others, charitable to all, modest in deportment, and aware of the transient significance of the individual relative to the scale of the universe and its doings. Spinoza bypassed Christianity but incorporated Christ in his system. In fact, he may have modeled the last phase of his life on Christ's. He seems to have melded Christ with the stoic streak of the *marrano* tradition, arriving at an ultimate joy by denying himself many smaller joys along the way.

The philosopher C. S. Peirce notes this connection clearly: "Spinoza's ideas are eminently ideas to affect human conduct. If, in accordance with the recommendation of Jesus, we are to judge of ethical doctrines and of philosophy in general by its practical fruits,

we cannot but consider Spinoza as a very weighty authority; for
probably no writer of modern times has so much determined men
towards an elevated mode of life. Although his doctrine contains
many things which are unchristian, yet they are unchristian rather
intellectually than practically. In part, at least, Spinozism is, after all,
a special development of Christianity; and the practical upshot of it
is decidedly more Christian than that of any current system of the-
ology." (Charles Sanders Peirce, "Spinoza's Ethic," *The Nation*, Vol.
LIX [1894]: 344–45.)

6   Jonathan Bennett, *A Study of Spinoza's Ethics* (Indianapolis, Ind.:
    Hackett Publishing Company, 1984).

7   See Barbara Stafford's *Devices of Wonder: From the World in a Box to
    Images on a Screen* (Los Angeles: Getty Research Institute, 2001).

8   Albert Einstein, *The World as I See It* (cited earlier).

9   Ibid.

10 Ibid.

11 Richard Warrington Baldwin Lewis, *The Jameses* (New York: Farrar,
    Straus and Giroux, 1991).

12 William James, *The Varieties of Religious Experience* (Cambridge,
    Mass.: Harvard University Press, 1985), Lecture I: *The Varieties of
    Religious Experience*.

13 William James, ibid., Lecture VI.

14 William James, ibid., Lecture VI.

15 For a clear statement on the shortcomings of such attempts see
    Jerome Groopman, "God on the Brain," *The New Yorker*, September
    17 (2001): 165–68.

16 I should add that there are, to be sure, many other kinds of spiritual
    experience and by no means do I wish to be restrictive. Some spiri-
    tual experiences can be described less as a feeling than as a form of
    mental clarity, of focused, selfless attention. In keeping with our
    discussion on mind and body relations, it is perhaps correct, how-
    ever, that most forms of spiritual experience require a particular
    configuration of the body, and depend in fact on the body being ac-
    tively placed in a certain mode.

17 Spinoza, *The Ethics* (cited earlier).

Glossary

Action potential: The all or none electrical pulse conducted along the neuron's axon from the cell body towards the multiple branches at the far end of the axon.

Axon: The typically singular output fiber of a neuron. One single axon can make contacts (synapses) with the dendrites of numerous other neurons and thus disseminate signals extensively.

Basal forebrain: A set of small nuclei located in front and beneath the basal ganglia. These nuclei are involved in the execution of regulatory behaviors, including emotions, and also play a role in learning and memory.

Brain stem: A set of small nuclei and white matter pathways placed between the diencephalon (the aggregate of the thalamus and hypothalamus) and the spinal cord. The nuclei in the brainstem are involved in the regulation of life, for example, the regulation of metabolism. The execution of emotions depends on many such nuclei. Extensive damage to nuclei in the upper and posterior part of the brainstem leads to loss of consciousness. The brainstem is a conduit for pathways from the brain to the body (carrying signals related to movement); and from the body to the brain (carrying the signals that inform the brain's body maps).

Central nervous system: The aggregate constituted by the cerebral hemispheres, the cerebellum, the diencephalon (which is formed by the thalamus and hypothalamus), the brain stem, and the spinal cord. See Appendix II, figure 1.

Cerebral cortex: The all-encompassing mantle that covers the cerebrum (the combination of the left and right cerebral hemispheres). The cortex covers the entire cerebral surfaces including those that are located in the depth of crevices that give the brain its characteristic folded appearance and are known as fissures and sulci. The cerebral cortex is organized in layers parallel to one another and to the brain's surface. The layers are not unlike those of a cake and are made of neurons. The neurons in the cerebral cortex receive signals from other neurons (in other regions of the cerebral cortex or elsewhere in the brain) and initiate signals toward other neurons in many other regions (inside and outside the cerebral cortex and not). The cerebral cortex has evolutionarily old components (for example, the so-called limbic cortices of which the cingulate region is a part) and evolutionarily modern components (known as the neocortex). The cellular architecture of the cortex varies from region to region and is easily identified by the numbers of Brodmann's map (see figure 2, Appendix II).

Cerebellum: A sort of minibrain located under the posterior part of the big brain (the cerebrum). As is the case with the cerebrum, the cerebellum has two hemispheres, left and right, and each hemisphere is covered by a cortex. The cerebellum is involved in the planning and execution of movements. It is indispensable for precise movements. There is good reason to believe, however, that the cerebellum is also involved in cognitive processes. Without a doubt, it plays a part in the execution and tuning of emotional responses.

Cerebrum: A virtual synonym of brain. It is formed by two large structures, the cerebral hemispheres, which occupy most of the intracranial cavity. Each cerebral hemisphere is entirely covered by the cerebral cortex.

Corpus callosum: A thick collection of axons connecting the neurons in the left and right hemispheres, in both directions, transversally.

CT: The initials CT stand for "computerized tomography" and are frequently used to designate "x-ray computerized tomography scans." CT was the first modern brain imaging technique (it appeared in 1973) and, although it has been overtaken by MR and PET, it remains a mainstay in the clinical neurological evaluation of conditions such as strokes.

Enzymes: Usually large protein molecules that serve as catalysts of biochemical reactions.

Gray matter: The darker sectors of the central nervous system are known as the "gray matter," while the pale sectors are known as "white matter." The gray matter corresponds to tightly packed collections of the neurons' cell bodies, while the white matter corresponds mostly to the neuron's axons, the usually singular fiber outputs of the neuron's cell body. The gray matter comes in two main varieties: the layered variety, found in the cortex of the cerebrum and cerebellum; and the nucleus variety, in which the neurons are organized like grapes in a bowl rather than in layers.

Lesion: An area of circumscribed damage to the central nervous system or to a peripheral nerve. It is usually caused by ischemia (a reduction or interruption of blood supply) or by mechanical injury. The normal neuroanatomical structure is destroyed in lesioned tissue.

MRI: The initials MRI stand for magnetic resonance imaging, also known as MR for short. MR is one of the fundamental methods of brain imaging. It can provide extremely refined images of brain structure as well as functional images of the type offered by PET. When it is used for functional imaging purposes it is usually designated as fMR or fMRI.

Neuron: The fundamental kind of nerve cell. Neurons come in many sizes and shapes but are usually formed by a cell body (the part of the neuron that gives the darker tone to the so-called gray matter) and by an output fiber known as the axon. In general, the neuron's input fibers are the dendrites, tree-like arborizations arising from the neuron's cell body. In addition to cell bodies, axons, and dendrites, the mass of the central nervous system is also formed by glial cells. Glial cells provide scaffolding for neurons and support their metabolism in a variety of ways. It is not entirely clear if glial cells also provide an additional signaling function.

Neurotransmitters and Neuromodulators: Molecules released by neurons that excite or inhibit the activity of other neurons (as glutamate and gamma-amino-butyric acid do), or modulate the activity of entire collections of neurons (as dopamine, serotonin, norepinephrine, and acetylcholine do).

Nuclei: A non-layered aggregate of neurons (see gray matter). Nuclei can be large or small. Large nuclei include the caudate, the putamen, and the pallidum, which together form the basal ganglia. Examples of small nuclei include those in the thalamus, hypothalamus, and brain stem. The amygdala is a fairly large aggregate of small nuclei hidden inside the temporal lobe.

Pathway: A collection of aligned axons carrying signals from one region to the other within the central nervous systems. It is the equivalent of a nerve in the peripheral nervous system. Also referred to as a "projection."

Periaqueductal gray: A collection of nuclei in the upper part of the brain stem involved in the execution of emotions.

Peripheral nervous system: The sum total of all the nerves that exit and enter the central nervous system.

PET scanning: The initials PET stand for positron emission tomography. This is one of the main techniques of functional imaging, and it permits the identification of brain region whose activity is either increased or reduced while the brain is engaged in performing a particular task.

Projection: See pathway.

Somatosensory: Having to do with sensory signaling from any part of the body (soma) to the central nervous system. The term interoceptive (see figure 3.5a) designates the part of body signalling hailing from the body's interior.

Substantia nigra: One of the small brain-stem nuclei that produces dopamine and delivers it to the brain structures located above it. Dopamine is essential for normal movement and is involved in reward.

Synapse: The microscopic region where the axon of one neuron connects with another neuron; for example, the region where the axon of one neuron connects with dendrites of another neuron. In essence, the synaptic connection is a gap rather than a bridge. The link is established by neurotransmitter molecules released on the axon side, as a result of the electrical impulse that traveled down the axon. The released molecules are taken up by receptors in the neuron they target and thus contribute to the activation of that neuron.

Acknowledgments

I begin by acknowledging the colleagues and friends who read this manuscript at different stages of its development, sometimes more than once, and gave me so many valuable criticisms and suggestions. I have no words to thank them for their generosity. They include Jean Pierre Changeux, David Hubel, Charles Rockland, Steven Nadler, Stuart Hampshire, Patricia Churchland, Paul Churchland, Thomas Metzinger, Oliver Sacks, Stefan Heck, Fernando Gil, David Rudrauf, Peter Sacks, Peter Brook, John Burnham Schwartz, and Jack Fromkin. They are not to be blamed for the oddities and errors that remain.

My colleagues at the University of Iowa and the Salk Institute were equally supportive, in particular Antoine Bechara, Ralph Adolphs, Daniel Tranel, and Josef Parvizi who also read the manuscript and made helpful suggestions. As always I thank the National Institute of Neurological Diseases and Stroke and the Mathers Foundation without whose support we could not have created, for scientists, patients, and students, the unique work atmosphere of the Division of Cognitive Neuroscience at the University of Iowa's Department of Neurology.

I must thank all those who helped me over the past five years with the various bibliographic searches required by the project: Maria de Sousa and José Horta, who found numerous old manuscripts on Spinoza in Portuguese libraries; Margaret Gullan-Whur, Maria Luisa Ribeiro Ferreira, and Diogo Pires Aurélio, three Spinoza scholars who patiently answered my questions on the great man; Mariana Anagnostopoulus, who located a key reference on the stoics for me; Thomas Casey, who clarified a few queries regarding the Boeing 777; and Arthur Bonfield, for a most helpful conversation on Thomas Jefferson and John Locke.

My assistant Neal Purdum coordinated the various parts of the manuscript with remarkable professionalism and evenness of temper;

he also typed most of it with the help of Betty Redeker, whose patience for working from my handwriting, after twenty years, is nothing if not miraculous. A million thanks to both for their dedication, and to Ken Manzel who helped promptly with library research as did Carol Devore, on several occasions.

Many heartfelt thanks to Donna Wares for her superb editing and to David Hough for making the book come together.

This book would not have been written without the enthusiasm and support of two long-time friends, Jane Isay and Michael Carlisle, and of Hanna Damasio, colleague, worst critic, best critic, and day-to-day source of inspiration and reason.

Index